ELECTRICAL REFERENCES

2014 EDITION

A NOTE FROM THE AUTHOR...

Ugly's Electrical References is designed to be used as a quick on-the-job reference in the electrical industry. Used worldwide by electricians, engineers, contractors, designers, maintenance workers, instructors, and the military, *Ugly's* contains the most commonly required information in an easy-to-read format.

Ugly's Electrical References is not intended to be a substitute for the *National Electrical Code*®.

We salute the National Fire Protection Association for their dedication to the protection of lives and property from fire and electrical hazards through the sponsorship of the *National Electrical Code*®.

JONES & BARTLETT
LEARNING

TABLE OF CONTENTS

TABLE OF CONTENTS (continued)

TABLE OF CONTENTS (continued)

TABLE OF CONTENTS (continued)

TABLE OF CONTENTS (continued)

TABLE OF CONTENTS (continued)

TABLE OF CONTENTS (continued)

TABLE OF CONTENTS (continued)

TABLE OF CONTENTS (continued)

OHM'S LAW

The rate of the flow of the current is equal to electromotive force divided by resistance.

I = Intensity of Current = Amperes

E = Electromotive Force = Volts

R = Resistance = Ohms

P = Power = Watts

The three basic Ohm's Law formulas are:

$$I = \frac{E}{R} \qquad R = \frac{E}{I} \qquad E = I \times R$$

Below is a chart containing the formulas related to Ohm's Law. To use the chart, from the center circle, select the value you need to find, I (amps), R (ohms), E (volts) or P (watts). Then select the formula containing the values you know from the corresponding chart quadrant.

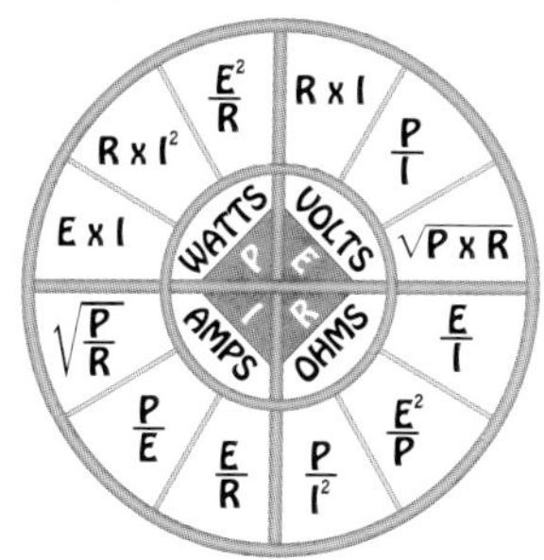

EXAMPLE:

An electrical appliance is rated at 1200 watts and is connected to 120 volts. How much current will it draw?

$$\text{Amperes} = \frac{\text{Watts}}{\text{Volts}} \qquad I = \frac{P}{E} \qquad I = \frac{1200}{120} = 10\ A$$

What is the resistance of the same appliance?

$$\text{Ohms} = \frac{\text{Volts}}{\text{Amperes}} \qquad R = \frac{E}{I} \qquad R = \frac{120}{10} = 12\ \Omega$$

OHM'S LAW

In the preceding example, we know the following values:

I = amps = 10 A	R = ohms = 12 Ω
E = volts = 120 V	P = watts = 1200 W

We can now see how the twelve formulas in the Ohm's Law chart can be applied.

$$\text{AMPS} = \sqrt{\frac{\text{WATTS}}{\text{OHMS}}} \qquad I = \sqrt{\frac{P}{R}} = \sqrt{\frac{1200}{12}} = \sqrt{100} = 10\text{ A}$$

$$\text{AMPS} = \frac{\text{WATTS}}{\text{VOLTS}} \qquad I = \frac{P}{E} = \frac{1200}{120} = 10\text{ A}$$

$$\text{AMPS} = \frac{\text{VOLTS}}{\text{OHMS}} \qquad I = \frac{E}{R} = \frac{120}{12} = 10\text{ A}$$

$$\text{WATTS} = \frac{\text{VOLTS}^2}{\text{OHMS}} \qquad P = \frac{E^2}{R} = \frac{120^2}{12} = \frac{14{,}400}{12} = 1200\text{ W}$$

$$\text{WATTS} = \text{VOLTS} \times \text{AMPS} \qquad P = E \times I = 120 \times 10 = 1200\text{ W}$$

$$\text{WATTS} = \text{AMPS}^2 \times \text{OHMS} \qquad P = I^2 \times R = 10^2 \times 12 = 1200\text{ W}$$

$$\text{VOLTS} = \sqrt{\text{WATTS} \times \text{OHMS}} \qquad E = \sqrt{P \times R} = \sqrt{1200 \times 12} = \sqrt{14{,}400} = 120\text{ V}$$

$$\text{VOLTS} = \text{AMPS} \times \text{OHMS} \qquad E = I \times R = 10 \times 12 = 120\text{ V}$$

$$\text{VOLTS} = \frac{\text{WATTS}}{\text{AMPS}} \qquad E = \frac{P}{I} = \frac{1200}{10} = 120\text{ V}$$

$$\text{OHMS} = \frac{\text{VOLTS}^2}{\text{WATTS}} \qquad R = \frac{E^2}{P} = \frac{120^2}{1200} = \frac{14{,}400}{1200} = 12\ \Omega$$

$$\text{OHMS} = \frac{\text{WATTS}}{\text{AMPS}^2} \qquad R = \frac{P}{I^2} = \frac{1200}{10^2} = 12\ \Omega$$

$$\text{OHMS} = \frac{\text{VOLTS}}{\text{AMPS}} \qquad R = \frac{E}{I} = \frac{120}{10} = 12\ \Omega$$

SERIES CIRCUITS

A SERIES CIRCUIT is a circuit that has only one path through which the electrons may flow.

RULE 1: **The total current in a series circuit is equal to the current in any other part of the circuit.**

TOTAL CURRENT $I_T = I_1 = I_2 = I_3$, **etc.**

RULE 2: **The total voltage in a series circuit is equal to the sum of the voltages across all parts of the circuit.**

TOTAL VOLTAGE $E_T = E_1 + E_2 + E_3$, **etc.**

RULE 3: **The total resistance of a series circuit is equal to the sum of the resistances of all the parts of the circuit.**

TOTAL RESISTANCE $R_T = R_1 + R_2 + R_3$, **etc.**

FORMULAS FROM OHM'S LAW

$$\text{AMPERES} = \frac{\text{VOLTS}}{\text{RESISTANCE}} \quad \text{OR} \quad I = \frac{E}{R}$$

$$\text{RESISTANCE} = \frac{\text{VOLTS}}{\text{AMPERES}} \quad \text{OR} \quad R = \frac{E}{I}$$

$$\text{VOLTS} = \text{AMPERES} \times \text{RESISTANCE} \quad \text{OR} \quad E = I \times R$$

EXAMPLE 1: Find the total voltage, total current, and total resistance of the following series circuit.

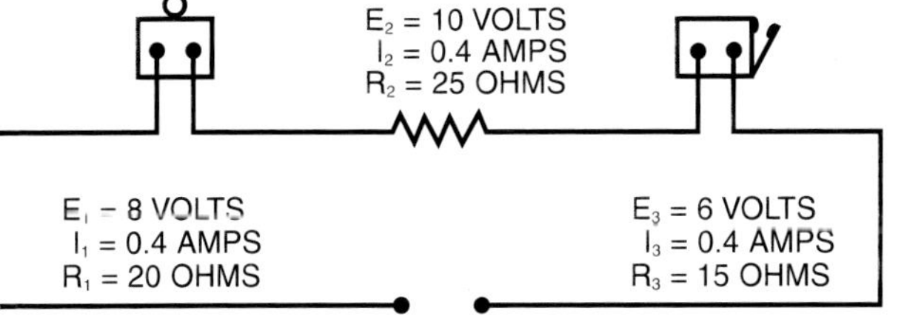

SERIES CIRCUITS

$E_T = E_1 + E_2 + E_3$
$= 8 + 10 + 6$
$E_T = 24$ VOLTS

$I_T = I_1 = I_2 = I_3$
$= 0.4 = 0.4 = 0.4$
$I_T = 0.4$ AMPS

$R_T = R_1 + R_2 + R_3$
$= 20 + 25 + 15$
$R_T = 60$ OHMS

EXAMPLE 2: Find E_T, E_1, E_3, I_T, I_1, I_2, I_4, R_T, R_2, and R_4. Remember that the total current in a series circuit is equal to the current in any other part of the circuit.

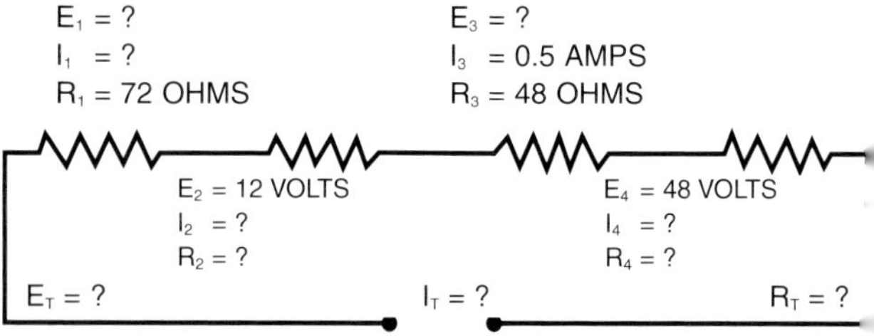

$E_1 = ?$
$I_1 = ?$
$R_1 = 72$ OHMS

$E_3 = ?$
$I_3 = 0.5$ AMPS
$R_3 = 48$ OHMS

$E_2 = 12$ VOLTS
$I_2 = ?$
$R_2 = ?$

$E_4 = 48$ VOLTS
$I_4 = ?$
$R_4 = ?$

$E_T = ?$ $I_T = ?$ $R_T = ?$

$I_T = I_1 = I_2 = I_3 = I_4$
$I_T = I_1 = I_2 = 0.5 = I_4$
$0.5 = 0.5 = 0.5 = 0.5 = 0.5$
$I_T = 0.5$ AMPS $I_2 = 0.5$ AMPS
$I_1 = 0.5$ AMPS $I_4 = 0.5$ AMPS

$E_T = E_1 + E_2 + E_3 + E_4$
$= 36 + 12 + 24 + 48$
$E_T = 120$ VOLTS

$R_T = R_1 + R_2 + R_3 + R_4$
$= 72 + 24 + 48 + 96$
$R_T = 240$ OHMS

$$R_2 = \frac{E_2}{I_2} = \frac{12}{0.5}$$
$R_2 = 24$ OHMS

$E_1 = I_1 \times R_1$
$= 0.5 \times 72$
$E_1 = 36$ VOLTS

$E_3 = I_3 \times R_3$
$= 0.5 \times 48$
$E_3 = 24$ VOLTS

$$R_4 = \frac{E_4}{I_4} = \frac{48}{0.5}$$
$R_4 = 96$ OHMS

PARALLEL CIRCUITS

A PARALLEL CIRCUIT is a circuit that has more than one path through which the electrons may flow.

RULE 1: The total current in a parallel circuit is equal to the sum of the currents in all the branches of the circuit.
TOTAL CURRENT $I_T = I_1 + I_2 + I_3$, etc.

RULE 2: The total voltage across any branch in parallel is equal to the voltage across any other branch and is also equal to the total voltage.
TOTAL VOLTAGE $E_T = E_1 = E_2 = E_3$, etc.

RULE 3: The total resistance of a parallel circuit is found by applying Ohm's Law to the total values of the circuit.

$$\text{TOTAL RESISTANCE} = \frac{\text{TOTAL VOLTAGE}}{\text{TOTAL AMPERES}} \quad \text{OR} \quad R_T = \frac{E_T}{I_T}$$

EXAMPLE 1: Find the total current, total voltage, and total resistance of the following parallel circuit.

E_1 = 120 VOLTS
I_1 = 2 AMPS
R_1 = 60 OHMS

E_2 = 120 VOLTS
I_2 = 1.5 AMPS
R_2 = 80 OHMS

E_3 = 120 VOLTS
I_3 = 1 AMPS
R_3 = 120 OHMS

$$\begin{aligned} I_T &= I_1 + I_2 + I_3 \\ &= 2 + 1.5 + 1 \\ I_T &= 4.5 \text{ AMPS} \end{aligned}$$

$$\begin{aligned} E_T &= E_1 = E_2 = E_3 \\ &= 120 = 120 = 120 \\ E_T &= 120 \text{ VOLTS} \end{aligned}$$

$$R_T = \frac{E_T}{I_T} = \frac{120 \text{ VOLTS}}{4.5 \text{ AMPS}} = 26.66 \text{ OHMS RESISTANCE}$$

NOTE. In a parallel circuit, the total resistance is always less than the resistance of any branch. If the branches of a parallel circuit have the same resistance, then each will draw the same current. If the branches of a parallel circuit have different resistances, then each will draw a different current. In either series or parallel circuits, the larger the resistance, the smaller the current drawn.

PARALLEL CIRCUITS

To determine the total resistance in a parallel circuit when the total current and total voltage are unknown:

$$\frac{1}{\text{TOTAL RESISTANCE}} = \frac{1}{R_1} + \frac{1}{R_2} + \frac{1}{R_3} + \ldots \text{ETC.}$$

EXAMPLE 2: Find the total resistance.

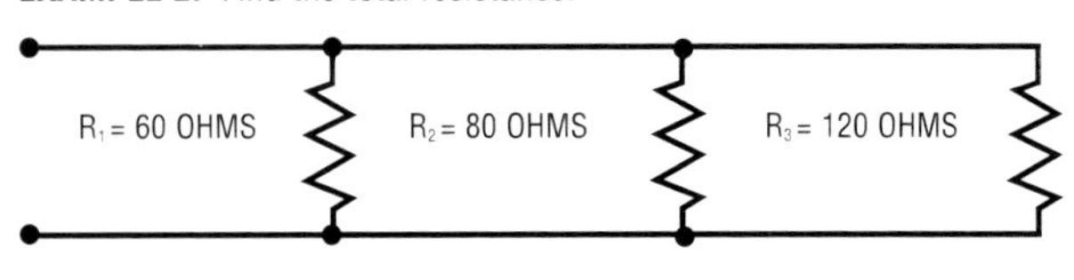

$$\frac{1}{R_T} = \frac{1}{R_1} + \frac{1}{R_2} + \frac{1}{R_3}$$

$$\frac{1}{R_T} = \frac{1}{60} + \frac{1}{80} + \frac{1}{120}$$

$$\frac{1}{R_T} = \frac{4 + 3 + 2}{240} = \frac{9}{240}$$ Use lowest common denominator (240).

$$\frac{1}{R_T} = \frac{9}{240}$$ Cross multiply.

For a review of Adding Fractions and Common Denominators, see Ugly's *pages 148–149.*

$9 \times R_T = 1 \times 240$ or $9R_T = 240$

Divide both sides of the equation by 9.

R_T = 26.66 OHMS RESISTANCE

NOTE: The total resistance of a number of EQUAL resistors in parallel is equal to the resistance of one resistor divided by the number of resistors.

$$\text{TOTAL RESISTANCE} = \frac{\text{RESISTANCE OF ONE RESISTOR}}{\text{NUMBER OF RESISTORS IN CIRCUIT}}$$

PARALLEL CIRCUITS

FORMULA: $$R_T = \frac{R}{N}$$

EXAMPLE 3: Find the total resistance.

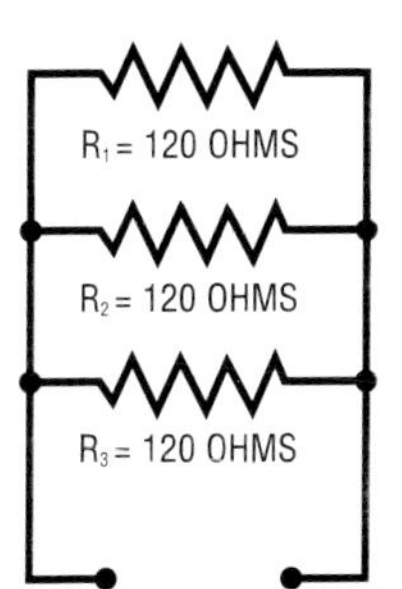

There are three resistors in parallel. Each has a value of 120 ohms resistance. According to the formula, if we divide the resistance of any one of the resistors by three, we will obtain the total resistance of the circuit.

$$R_T = \frac{R}{N} \quad \text{OR} \quad R_T = \frac{120}{3}$$

TOTAL RESISTANCE = 40 OHMS

NOTE: To find the total resistance of only two resistors in parallel, multiply the resistances, and then divide the product by the sum of the resistors.

FORMULA: TOTAL RESISTANCE $$= \frac{R_1 \times R_2}{R_1 + R_2}$$

EXAMPLE 4: Find the total resistance.

R_1 = 40 OHMS

R_2 = 80 OHMS

$$R_T = \frac{R_1 \times R_2}{R_1 + R_2}$$

$$= \frac{40 \times 80}{40 + 80}$$

$$R_T = \frac{3200}{120} = 26.66 \text{ OHMS}$$

COMBINATION CIRCUITS

In combination circuits, we combine series circuits with parallel circuits. Combination circuits make it possible to obtain the different voltages of series circuits and the different currents of parallel circuits.

EXAMPLE 1: PARALLEL-SERIES CIRCUIT.
Solve for all missing values.

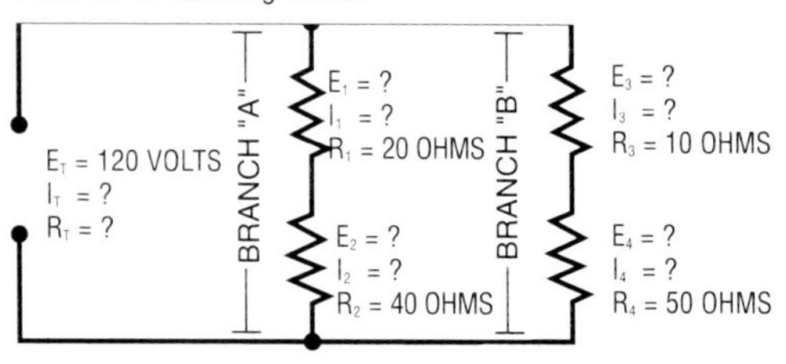

TO SOLVE:

1. Find the total resistance of each branch. Both branches are simple series circuits, so:

   $\mathbf{R_1 + R_2 = R_A}$
   20 + 40 = 60 ohms total resistance of branch "A"

   $\mathbf{R_3 + R_4 = R_B}$
   10 + 50 = 60 ohms total resistance of branch "B"

2. Redraw the circuit, combining resistors ($R_1 + R_2$) and ($R_3 + R_4$) so that each branch will have only one resistor.

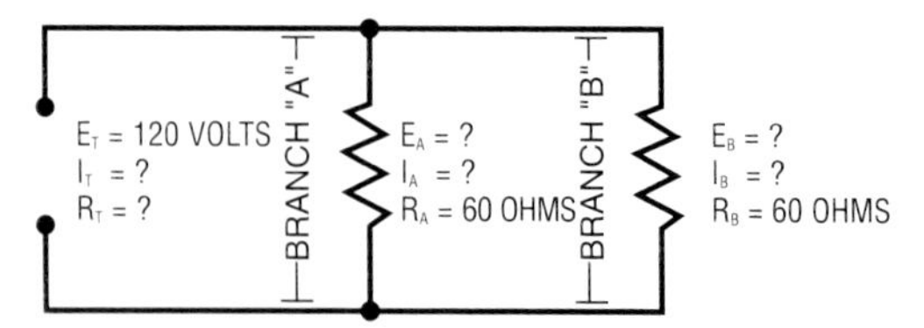

COMBINATION CIRCUITS

NOTE: We now have a simple parallel circuit, so:

$$E_T = E_A = E_B$$
$$120\text{ V} = 120\text{ V} = 120\text{ V}$$

We now have a parallel circuit with only two resistors, and they are of equal value. We have a choice of three different formulas that can be used to solve for the total resistance of the circuit.

(1) $$R_T = \frac{R_A \times R_B}{R_A + R_B} = \frac{60 \times 60}{60 + 60} = \frac{3600}{120} = 30\text{ OHMS}$$

(2) When the resistors of a parallel circuit are of equal value,

$$R_T = \frac{R}{N} = \frac{60}{2} = 30\text{ OHMS} \qquad \text{OR}$$

(3) $$\frac{1}{R_T} = \frac{1}{R_A} + \frac{1}{R_B} = \frac{1}{60} + \frac{1}{60} = \frac{2}{60} = \frac{1}{30}$$

$$\frac{1}{R_T} = \frac{1}{30} \quad \text{OR} \quad 1 \times R_T = 1 \times 30 \quad \text{OR} \quad R_T = 30\text{ OHMS}$$

3. We know the values of E_T, R_T, E_A, R_A, E_B, R_B, R_1, R_2, R_3, and R_4. Next we will solve for I_T, I_A, I_B, I_1, I_2, I_3, and I_4.

$$I_T = \frac{E_T}{R_T} \quad \text{OR} \quad \frac{120}{30} = 4 \qquad I_T = 4\text{ AMPS}$$

$$I_A = \frac{E_A}{R_A} \quad \text{OR} \quad \frac{120}{60} = 2 \qquad I_A = 2\text{ AMPS}$$

$$I_A = I_1 = I_2 \quad \text{OR} \quad 2 = 2 = 2 \qquad I_1 = 2\text{ AMPS}, \; I_2 = 2\text{ AMPS}$$

$$I_B = \frac{E_B}{R_B} = \quad \text{OR} \quad \frac{120}{60} = 2 \qquad I_B = 2\text{ AMPS}$$

$$I_B = I_3 = I_4 \quad \text{OR} \quad 2 = 2 = 2 \qquad I_3 = 2\text{ AMPS}, \; I_4 = 2\text{ AMPS}$$

COMBINATION CIRCUITS

4. We know that resistors #1 and #2 of branch "A" are in series. We know too that resistors #3 and #4 of branch "B" are in series. We have determined that the total current of branch "A" is 2 A, and the total current of branch "B" is 2 A. By using the series formula, we can solve for the current of each branch.

BRANCH "A"	BRANCH "B"
$\mathbf{I_A = I_1 = I_2}$	$\mathbf{I_B = I_3 = I_4}$
$2 = 2 = 2$	$2 = 2 = 2$
I_1 = 2 AMPS	I_3 = 2 AMPS
I_2 = 2 AMPS	I_4 = 2 AMPS

5. We were given the resistance values of all resistors. R_1 = 20 OHMS, R_2 = 40 OHMS, R_3 = 10 OHMS, and R_4 = 50 OHMS. By using Ohm's Law, we can determine the voltage drop across each resistor.

$$\mathbf{E_1 = R_1 \times I_1}$$
$$= 20 \times 2$$
$$E_1 = 40 \text{ VOLTS}$$

$$\mathbf{E_3 = R_3 \times I_3}$$
$$= 10 \times 2$$
$$E_3 = 20 \text{ VOLTS}$$

$$\mathbf{E_2 = R_2 \times I_2}$$
$$= 40 \times 2$$
$$E_2 = 80 \text{ VOLTS}$$

$$\mathbf{E_4 = R_4 \times I_4}$$
$$= 50 \times 2$$
$$E_4 = 100 \text{ VOLTS}$$

EXAMPLE 2: SERIES PARALLEL CIRCUIT.
Solve for all missing values.

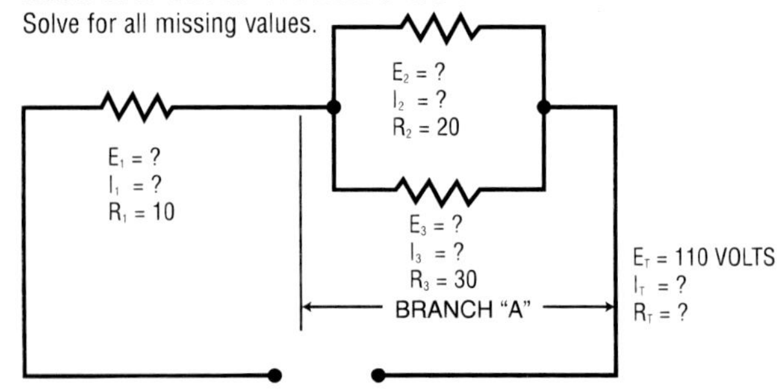

COMBINATION CIRCUITS

To solve:

1. We can see that resistors #2 and #3 are in parallel, and combined they are branch "A." When there are only two resistors, we use the following formula:

$$\mathbf{R_A} = \frac{\mathbf{R_2 \times R_3}}{\mathbf{R_2 + R_3}} \text{ OR } \frac{20 \times 30}{20 + 30} \text{ OR } \frac{600}{50} \text{ OR 12 OHMS}$$

2. We can now redraw our circuit as a simple series circuit.

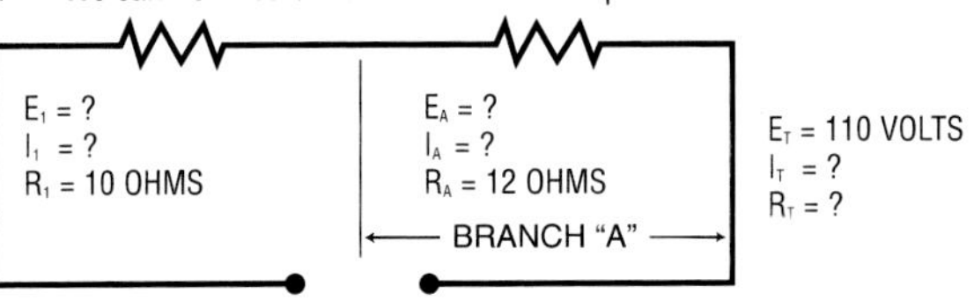

3. In a series circuit,
   $\mathbf{R_T = R_1 + R_A}$ OR $R_T = 10 + 12$ OR 22 OHMS
   By using Ohm's Law,

$$\mathbf{I_T} = \frac{\mathbf{E_T}}{\mathbf{R_T}} = \frac{110}{22} = 5 \text{ AMPS}$$

   In a series circuit,
   $\mathbf{I_T = I_1 = I_A}$ OR I_T = 5 AMPS, I_1 = 5 AMPS, and I_A = 5 AMPS

   By using Ohm's Law,
   $\mathbf{E_1 = I_1 \times R_1} = 5 \times 10 = 50$ VOLTS
   $E_T - E_1 = E_A$ OR $110 - 50 = 60$ VOLTS $= E_A$

   In a parallel circuit,
   $\mathbf{E_A = E_2 = E_3}$ OR E_A = 60 VOLTS
   E_2 = 60 VOLTS, and E_3 = 60 VOLTS

   By using Ohm's Law,

$$\mathbf{I_2} = \frac{\mathbf{E_2}}{\mathbf{R_2}} = \frac{60}{20} = 3 \text{ AMPS}$$

$$\mathbf{I_3} = \frac{\mathbf{E_3}}{\mathbf{R_3}} = \frac{60}{30} = 2 \text{ AMPS}$$

COMBINATION CIRCUITS

PROBLEM:

Solve for total resistance.
Redraw circuit as many times as necessary.
Correct answer is 100 ohms.

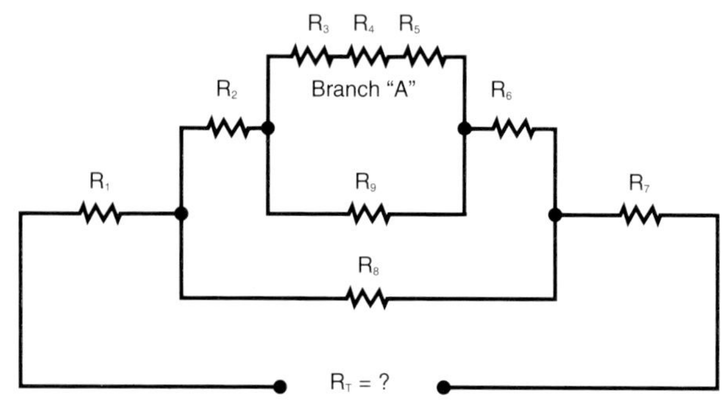

GIVEN VALUES:

R_1 = 15 OHMS
R_2 = 35 OHMS
R_3 = 50 OHMS
R_4 = 40 OHMS
R_5 = 30 OHMS
R_6 = 25 OHMS
R_7 = 10 OHMS
R_8 = 300 OHMS
R_9 = 60 OHMS

COMMON ELECTRICAL DISTRIBUTION SYSTEMS

120/240-Volt, Single-Phase, Three-Wire System

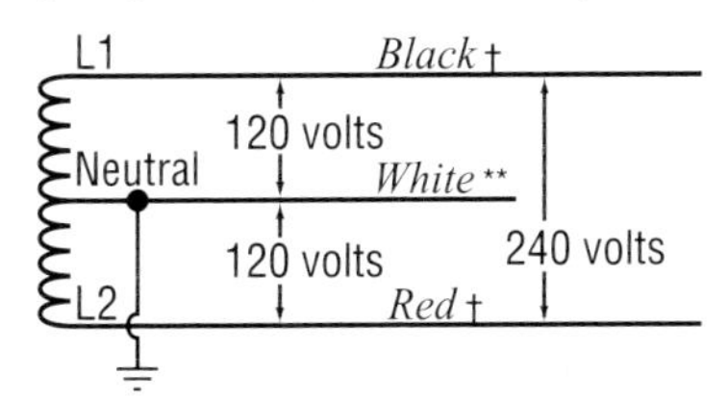

† • **Line one** ungrounded conductor colored **black**
† • **Line two** ungrounded conductor colored **red**
• Grounded neutral conductor colored **white or gray

120/240-Volt, Three-Phase, Four-Wire System (Delta High Leg)

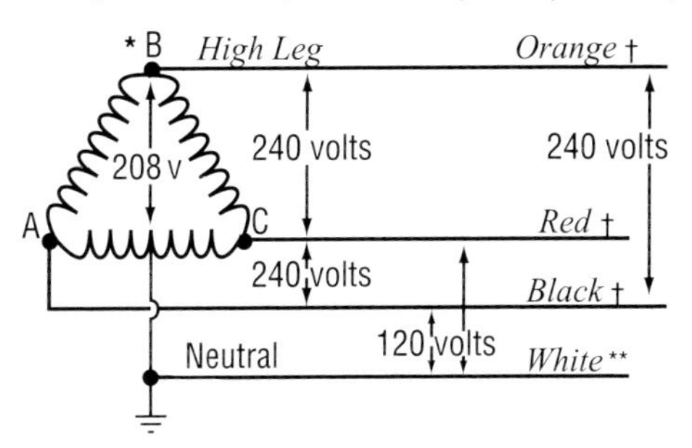

† • **A** phase ungrounded conductor colored **black**
†* • **B** phase ungrounded conductor colored **orange** or tagged (high leg). (Caution: 208 V orange to white)
† • **C** phase ungrounded conductor colored **red**
** • Grounded conductor colored **white** or gray (center tap)

** Grounded conductors are required to be white or gray or three white or gray stripes. See *NEC* 200.6(A).
* B phase of delta high leg must be orange or tagged.
† Ungrounded conductor colors may be other than shown; see local ordinances or specifications.

COMMON ELECTRICAL DISTRIBUTION SYSTEMS

120/208-Volt, Three-Phase, Four-Wire System (Wye Connected)

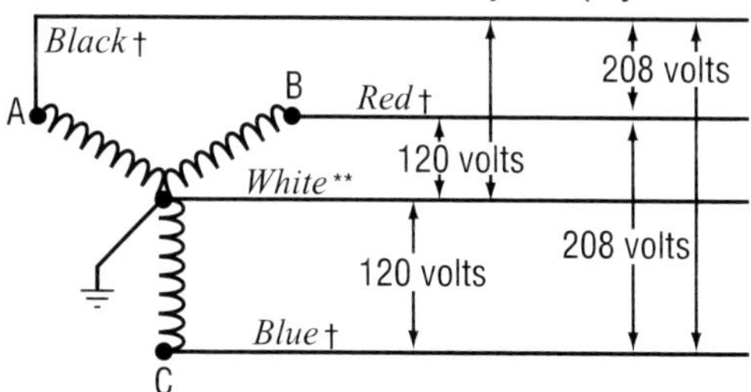

† • **A** phase ungrounded conductor colored **black**
† • **B** phase ungrounded conductor colored **red**
† • **C** phase ungrounded conductor colored **blue**
• Grounded neutral conductor colored **white or gray

277/480-Volt, Three-Phase, Four-Wire System (Wye Connected)

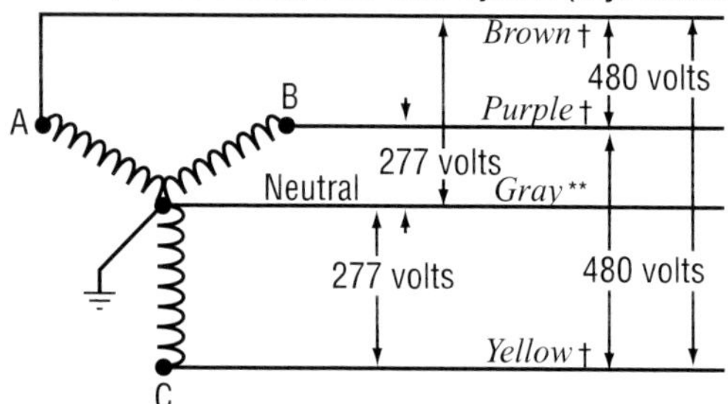

† • **A** phase ungrounded conductor colored **brown**
† • **B** phase ungrounded conductor colored **purple**
† • **C** phase ungrounded conductor colored **yellow**
• Grounded neutral conductor colored **gray

** Grounded conductors are required to be white or gray or three white or gray stripes. See *NEC* 200.6(A).
* B phase of delta high leg must be orange or tagged.
† Ungrounded conductor colors may be other than shown; see local ordinances or specifications.

ELECTRICAL FORMULAS FOR CALCULATING AMPERES, HORSEPOWER, KILOWATTS, AND KVA

TO FIND	DIRECT CURRENT	ALTERNATING CURRENT		
		SINGLE PHASE	TWO PHASE/FOUR WIRE	THREE PHASE
AMPERES WHEN "HP" IS KNOWN	$\frac{HP \times 746}{E \times \%EFF}$	$\frac{HP \times 746}{E \times \%EFF \times PF}$	$\frac{HP \times 746}{E \times \%EFF \times PF \times 2}$	$\frac{HP \times 746}{E \times \%EFF \times PF \times 1.73}$
AMPERES WHEN "KW" IS KNOWN	$\frac{KW \times 1000}{E}$	$\frac{KW \times 1000}{E \times PF}$	$\frac{KW \times 1000}{E \times PF \times 2}$	$\frac{KW \times 1000}{E \times PF \times 1.73}$
AMPERES WHEN "KVA" IS KNOWN		$\frac{KVA \times 1000}{E}$	$\frac{KVA \times 1000}{E \times 2}$	$\frac{KVA \times 1000}{E \times 1.73}$
KILOWATTS (True Power)	$\frac{E \times I}{1000}$	$\frac{E \times I \times PF}{1000}$	$\frac{E \times I \times PF \times 2}{1000}$	$\frac{E \times I \times PF \times 1.73}{1000}$
KILOVOLT-AMPERES "KVA" (Apparent Power)		$\frac{E \times I}{1000}$	$\frac{E \times I \times 2}{1000}$	$\frac{E \times I \times 1.73}{1000}$
HORSEPOWER	$\frac{E \times I \times \%EFF}{746}$	$\frac{E \times I \times \%EFF \times PF}{746}$	$\frac{E \times I \times \%EFF \times PF \times 2}{746}$	$\frac{E \times I \times \%EFF \times PF \times 1.73}{746}$

PERCENT EFFICIENCY = % EFF = $\frac{\text{OUTPUT (WATTS)}}{\text{INPUT (WATTS)}}$ POWER FACTOR = PF = $\frac{\text{POWER USED (WATTS)}}{\text{APPARENT POWER}} = \frac{KW}{KVA}$

NOTE: DIRECT-CURRENT FORMULAS DO NOT USE (PF, 2, OR 1.73).
SINGLE-PHASE FORMULAS DO NOT USE (2 OR 1.73).
TWO-PHASE/FOUR-WIRE FORMULAS DO NOT USE (1.73).
THREE-PHASE FORMULAS DO NOT USE (2).

E = VOLTS
I = AMPERES
W = WATTS

TO FIND AMPERES

Direct Current:

A. When *HORSEPOWER* is known:

$$\text{AMPERES} = \frac{\text{HORSEPOWER} \times 746}{\text{VOLTS} \times \text{EFFICIENCY}} \quad \text{or} \quad I = \frac{\text{HP} \times 746}{E \times \%\text{EFF}}$$

What current will a travel-trailer toilet draw when equipped with a 12-V, ⅛-HP motor, having a 96% efficiency rating?

$$I = \frac{\text{HP} \times 746}{E \times \%\text{EFF}} = \frac{746 \times \frac{1}{8}}{12 \times 0.96} = \frac{93.25}{11.52} = 8.09 \text{ AMPS}$$

B. When *KILOWATTS* are known:

$$\text{AMPERES} = \frac{\text{KILOWATTS} \times 1000}{\text{VOLTS}} \quad \text{or} \quad I = \frac{\text{KW} \times 1000}{E}$$

A 75-KW, 240-V direct-current generator is used to power a variable-speed conveyor belt at a rock-crushing plant. Determine the current.

$$I = \frac{\text{KW} \times 1000}{E} = \frac{75 \times 1000}{240} = 312.5 \text{ AMPS}$$

Single Phase:

A. When *WATTS, VOLTS, and POWER FACTOR* are known:

$$\text{AMPERES} = \frac{\text{WATTS}}{\text{VOLTS} \times \text{POWER FACTOR}} \quad \text{or} \quad \frac{P}{E \times \text{PF}}$$

Determine the current when a circuit has a 1500-watt load, a power factor of 86%, and operates from a single-phase, 230-volt source.

$$I = \frac{1500}{230 \times 0.86} = \frac{1500}{197.8} = 7.58 \text{ AMPS}$$

TO FIND AMPERES

Single Phase (continued):

B. When *HORSEPOWER* is known:

$$\textbf{AMPERES} = \frac{\textbf{HORSEPOWER x 746}}{\textbf{VOLTS x EFFICIENCY x POWER FACTOR}}$$

Determine the amp-load of a single-phase, ½-HP, 115-V motor. The motor has an efficiency rating of 92% and a power factor of 80%.

$$I = \frac{HP \times 746}{E \times \%EFF \times PF} = \frac{\frac{1}{2} \times 746}{115 \times 0.92 \times 0.80} = \frac{373}{84.64}$$

$$I = 4.4 \text{ AMPS}$$

C. When *KILOWATTS* are known:

$$\textbf{AMPERES} = \frac{\textbf{KILOWATTS x 1000}}{\textbf{VOLTS x POWER FACTOR}} \quad \text{or} \quad I = \frac{KW \times 1000}{E \times PF}$$

A 230-V, single-phase circuit has a 12-KW power load, and operates at 84% power factor. Determine the current.

$$I = \frac{KW \times 1000}{E \times PF} = \frac{12 \times 1000}{230 \times 0.84} = \frac{12{,}000}{193.2} = 62 \text{ A}$$

D. When *KILOVOLT-AMPERE* is known:

$$\textbf{AMPERES} = \frac{\textbf{KILOVOLT-AMPERE x 1000}}{\textbf{VOLTS}} \quad \text{or} \quad I = \frac{KVA \times 1000}{E}$$

A 115-V, 2-KVA, single-phase generator operating at full load will deliver 17.4 A. (Prove.)

$$I = \frac{2 \times 1000}{115} = \frac{2000}{115} = 17.4 \text{ A}$$

REMEMBER: By definition, amperes is the rate of the flow of the current.

TO FIND AMPERES

Three Phase:

A. When *WATTS, VOLTS, and POWER FACTOR* are known:

$$\text{AMPERES} = \frac{\text{WATTS}}{\text{VOLTS} \times \text{POWER FACTOR} \times 1.73}$$

or

$$I = \frac{P}{E \times PF \times 1.73}$$

Determine the current when a circuit has a 1500-watt load, a power factor of 86%, and operates from a three-phase, 230-volt source.

$$I = \frac{P}{E \times PF \times 1.73} = \frac{1500}{230 \times 0.86 \times 1.73} = \frac{1500}{342.2}$$

$$I = 4.4 \text{ AMPS}$$

B. When *HORSEPOWER* is known:

$$\text{AMPERES} = \frac{\text{HORSEPOWER} \times 746}{\text{VOLTS} \times \text{EFFICIENCY} \times \text{POWER FACTOR} \times 1.73}$$

OR

$$I = \frac{HP \times 746}{E \times \%EFF \times PF \times 1.73}$$

Determine the amp-load of a three-phase, ½-HP, 230-V motor. The motor has an efficiency rating of 92% and a power factor of 80%.

$$I = \frac{HP \times 746}{E \times \%EFF \times PF \times 1.73} = \frac{\frac{1}{2} \times 746}{230 \times .92 \times .80 \times 1.73}$$

$$= \frac{373}{293} = 1.27 \text{ AMPS}$$

TO FIND AMPERES

Three Phase (continued):

C. When *KILOWATTS* are known:

$$\text{AMPERES} = \frac{\text{KILOWATTS} \times 1000}{\text{VOLTS} \times \text{POWER FACTOR} \times 1.73}$$

$$\text{OR} \qquad I = \frac{\text{KW} \times 1000}{E \times PF \times 1.73}$$

A 230-V, three-phase circuit, has a 12-KW power load and operates at 84% power factor. Determine the current.

$$I = \frac{\text{KW} \times 1000}{E \times PF \times 1.73} = \frac{12 \times 1000}{230 \times 0.84 \times 1.73} = \frac{12{,}000}{334.24}$$

$$I = 36 \text{ AMPS}$$

D. When *KILOVOLT-AMPERE* is known:

$$\text{AMPERES} = \frac{\text{KILOVOLT-AMPERE} \times 1000}{E \times 1.73} = \frac{\text{KVA} \times 1000}{E \times 1.73}$$

A 230-V, 4-KVA, three-phase generator operating at full load will deliver 10 A. (Prove.)

$$I = \frac{\text{KVA} \times 1000}{E \times 1.73} = \frac{4 \times 1000}{230 \times 1.73} = \frac{4000}{397.9}$$

$$I = 10 \text{ AMPS}$$

TO FIND HORSEPOWER

Direct Current:

$$\text{HORSEPOWER} = \frac{\text{VOLTS x AMPERES x EFFICIENCY}}{746}$$

A 12-volt motor draws a current of 8.09 amperes and has an efficiency rating of 96%. Determine the horsepower.

$$HP = \frac{E \times I \times \%EFF}{746} = \frac{12 \times 8.09 \times 0.96}{746} = \frac{93.19}{746}$$

$$HP = 0.1249 = \tfrac{1}{8}\ HP$$

Single Phase:

$$HP = \frac{\text{VOLTS x AMPERES x EFFICIENCY x POWER FACTOR}}{746}$$

A single-phase, 115-volt ac motor has an efficiency rating of 92% and a power factor of 80%. Determine the horsepower if the amp-load is 4.4 amperes.

$$HP = \frac{E \times I \times \%EFF \times PF}{746} = \frac{115 \times 4.4 \times 0.92 \times 0.80}{746}$$

$$HP = \frac{372.416}{746} = 0.4992 = \tfrac{1}{2}\ HP$$

Three Phase:

$$HP = \frac{\text{VOLTS x AMPERES x EFFICIENCY x POWER FACTOR x 1.73}}{746}$$

A three-phase, 460-volt motor draws a current of 52 amperes. The motor has an efficiency rating of 94% and a power factor of 80%. Determine the horsepower.

$$HP = \frac{E \times I \times \%EFF \times PF \times 1.73}{746} = \frac{460 \times 52 \times 0.94 \times 0.80 \times 1.73}{746}$$

$$HP = 41.7\ HP$$

TO FIND WATTS

The electrical power in any part of a circuit is equal to the current in that part multiplied by the voltage across that part of the circuit.

A watt is the power used when 1 volt causes 1 ampere to flow in a circuit.

One horsepower is the amount of energy required to lift 33,000 pounds, 1 foot, in 1 minute. The electrical equivalent of 1 horsepower is 745.6 watts. One watt is the amount of energy required to lift 44.26 pounds, 1 foot, in 1 minute. Watts is power, and power is the amount of work done in a given time.

When *VOLTS AND AMPERES* are known:

POWER (WATTS) = VOLTS x AMPERES

A 120-volt ac circuit draws a current of 5 amperes. Determine the power consumption.

$\mathbf{P = E \times I} = 120 \times 5 = 600$ WATTS

We can now determine the resistance of this circuit.

POWER = RESISTANCE x $(\text{AMPERES})^2$

$P = R \times I^2$ OR $600 = R \times 25$

Divide both sides of the equation by 25.

$\frac{600}{25} = R$ OR $R = 24$ OHMS

or

$$\textbf{POWER} = \frac{(\textbf{VOLTS})^2}{\textbf{RESISTANCE}} \quad \textbf{OR} \quad \mathbf{P} = \frac{\mathbf{E^2}}{\mathbf{R}} \quad \text{OR} \quad 600 = \frac{120^2}{R}$$

$$R \times 600 = 120^2 \quad \text{OR} \quad R = \frac{14{,}400}{600} = 24 \text{ OHMS}$$

NOTE: Refer to the formulas of the Ohm's Law chart on page 1.

TO FIND KILOWATTS

Direct Current:

$$\text{KILOWATTS} = \frac{\text{VOLTS} \times \text{AMPERES}}{1000}$$

A 120-volt dc motor draws a current of 40 amperes. Determine the kilowatts.

$$\text{KW} = \frac{E \times I}{1000} = \frac{120 \times 40}{1000} = \frac{4800}{1000} = 4.8 \text{ KW}$$

Single Phase:

$$\text{KILOWATTS} = \frac{\text{VOLTS} \times \text{AMPERES} \times \text{POWER FACTOR}}{1000}$$

A single-phase, 115-volt ac motor draws a current of 20 amperes and has a power factor rating of 86%. Determine the kilowatts.

$$\text{KW} = \frac{E \times I \times PF}{1000} = \frac{115 \times 20 \times 0.86}{1000} = \frac{1978}{1000} = 1.978 = 2 \text{ KW}$$

Three Phase:

$$\text{KILOWATTS} = \frac{\text{VOLTS} \times \text{AMPERES} \times \text{POWER FACTOR} \times 1.73}{1000}$$

A three-phase, 460-volt ac motor draws a current of 52 amperes and has a power factor rating of 80%. Determine the kilowatts.

$$\text{KW} = \frac{E \times I \times PF \times 1.73}{1000} = \frac{460 \times 52 \times 0.80 \times 1.73}{1000}$$

$$= \frac{33{,}105}{1000} = 33.105 = 33 \text{ KW}$$

TO FIND KILOVOLT-AMPERES

Single Phase:

$$\text{KILOVOLT-AMPERES} = \frac{\text{VOLTS} \times \text{AMPERES}}{1000}$$

A single-phase, 240-volt generator delivers 41.66 amperes at full load. Determine the kilovolt-amperes rating.

$$\text{KVA} = \frac{E \times I}{1000} = \frac{240 \times 41.66}{1000} = \frac{10{,}000}{1000} = 10\ \text{KVA}$$

Three Phase:

$$\text{KILOVOLT-AMPERES} = \frac{\text{VOLTS} \times \text{AMPERES} \times 1.73}{1000}$$

A three-phase, 460-volt generator delivers 52 amperes. Determine the kilovolt-amperes rating.

$$\text{KVA} = \frac{E \times I \times 1.73}{1000} = \frac{460 \times 52 \times 1.73}{1000} = \frac{41{,}382}{1000}$$

$$= 41.382 = 41\ \text{KVA}$$

NOTE: KVA = Apparent Power = Power Before Used, such as the rating of a transformer.

Kirchhoff's Laws

FIRST LAW (CURRENT):

The sum of the currents arriving at any point in a circuit must equal the sum of the currents leaving that point.

SECOND LAW (VOLTAGE):

The total voltage applied to any closed circuit path is always equal to the sum of the voltage drops in that path.

OR

The algebraic sum of all the voltages encountered in any loop equals zero.

TO FIND CAPACITANCE

Capacitance (C):

$$C = \frac{Q}{E} \quad \text{or} \quad \text{CAPACITANCE} = \frac{\text{COULOMBS}}{\text{VOLTS}}$$

Capacitance is the property of a circuit or body that permits it to store an electrical charge equal to the accumulated charge divided by the voltage. Expressed in farads.

A. To determine the total capacity of capacitors and/or condensers connected in series:

$$\frac{1}{C_T} = \frac{1}{C_1} + \frac{1}{C_2} + \frac{1}{C_3} + \frac{1}{C_4}$$

Determine the total capacity of four 12-microfarad capacitors connected in series.

$$\frac{1}{C_T} = \frac{1}{C_1} + \frac{1}{C_2} + \frac{1}{C_3} + \frac{1}{C_4}$$

$$= \frac{1}{12} + \frac{1}{12} + \frac{1}{12} + \frac{1}{12} = \frac{4}{12}$$

$$\frac{1}{C_T} = \frac{4}{12} \text{ OR } C_T \times 4 = 12 \text{ OR } C_T = \frac{12}{4} = 3 \text{ microfarads}$$

B. To determine the total capacity of capacitors and/or condensers connected in parallel:

$$C_T = C_1 + C_2 + C_3 + C_4$$

Determine the total capacity of four 12-microfarad capacitors connected in parallel:

$$C_T = C_1 + C_2 + C_3 + C_4$$

$$C_T = 12 + 12 + 12 + 12 = 48 \text{ microfarads}$$

A farad is the unit of capacitance of a condenser that retains 1 coulomb of charge with 1 volt difference of potential.

1 farad = 1,000,000 microfarads

SIX-DOT COLOR CODE FOR MICA AND MOLDED PAPER CAPACITORS

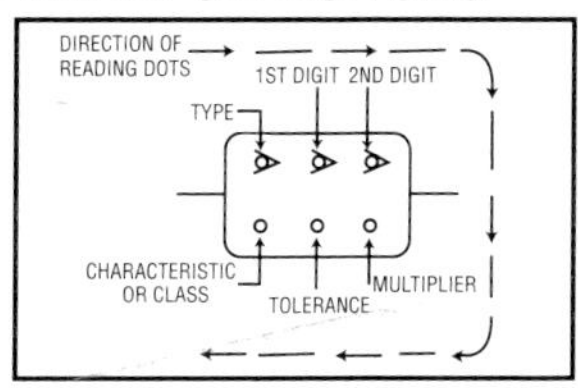

TYPE	COLOR	1ST DIGIT	2ND DIGIT	MULTIPLIER	TOLERANCE (%)	CHARACTERISTIC OR CLASS
JAN, MICA	BLACK	0	0	1	± 1	
	BROWN	1	1	10	± 2	
	RED	2	2	100	± 3	
	ORANGE	3	3	1000	± 4	
	YELLOW	4	4	10,000	± 5	APPLIES TO TEMPERATURE COEFFICIENT OR METHODS OF TESTING
	GREEN	5	5	100,000	± 6	
	BLUE	6	6	1,000,000	± 7	
	VIOLET	7	7	10,000,000	± 8	
	GRAY	8	8	100,000,000	± 9	
ETA, MICA	WHITE	9	9	1,000,000,000		
	GOLD			0.1	± 10	
MOLDED PAPER	SILVER			0.01	± 20	
	BODY					

RESISTOR COLOR CODE

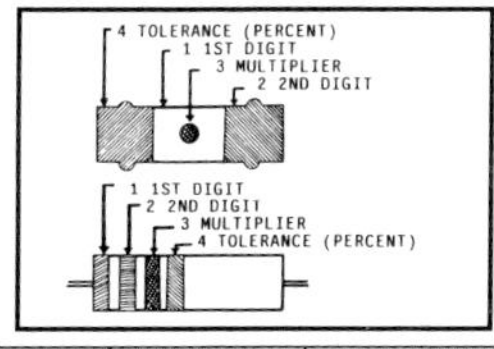

COLOR	1ST DIGIT	2ND DIGIT	MULTIPLIER	TOLERANCE (%)
BLACK	0	0	1	
BROWN	1	1	10	
RED	2	2	100	
ORANGE	3	3	1000	
YELLOW	4	4	10,000	
GREEN	5	5	100,000	
BLUE	6	6	1,000,000	
VIOLET	7	7	10,000,000	
GRAY	8	8	100,000,000	
WHITE	9	9	1,000,000,000	
GOLD			0.1	± 5%
SILVER			0.01	± 10%
NO COLOR				± 20%

MAXIMUM PERMISSIBLE CAPACITOR KVAR FOR USE WITH OPEN-TYPE, THREE-PHASE, 60-CYCLE INDUCTION MOTORS

	3600 RPM		1800 RPM		1200 RPM	
MOTOR RATING HP	MAXIMUM CAPACITOR RATING KVAR	REDUCTION IN LINE CURRENT %	MAXIMUM CAPACITOR RATING KVAR	REDUCTION IN LINE CURRENT %	MAXIMUM CAPACITOR RATING KVAR	REDUCTION IN LINE CURRENT %
10	3	10	3	11	3.5	14
15	4	9	4	10	5	13
20	5	9	5	10	6.5	12
25	6	9	6	10	7.5	11
30	7	8	7	9	9	11
40	9	8	9	9	11	10
50	12	8	11	9	13	10
60	14	8	14	8	15	10
75	17	8	16	8	18	10
100	22	8	21	8	25	9
125	27	8	26	8	30	9
150	32.5	8	30	8	35	9
200	40	8	37.5	8	42.5	9

	900 RPM		720 RPM		600 RPM	
MOTOR RATING HP	MAXIMUM CAPACITOR RATING KVAR	REDUCTION IN LINE CURRENT %	MAXIMUM CAPACITOR RATING KVAR	REDUCTION IN LINE CURRENT %	MAXIMUM CAPACITOR RATING KVAR	REDUCTION IN LINE CURRENT %
10	5	21	6.5	27	7.5	31
15	6.5	18	8	23	9.5	27
20	7.5	16	9	21	12	25
25	9	15	11	20	14	23
30	10	14	12	18	16	22
40	12	13	15	16	20	20
50	15	12	19	15	24	19
60	18	11	22	15	27	19
75	21	10	26	14	32.5	18
100	27	10	32.5	13	40	17
125	32.5	10	40	13	47.5	16
150	37.5	10	47.5	12	52.5	15
200	47.5	10	60	12	65	14

NOTE: If capacitors of a lower rating than the values given in the table are used, the percentage reduction in line current given in the table should be reduced proportionately.

POWER FACTOR CORRECTION

Table Values × KW of Capacitors Needed to Correct from Existing to Desired Power Factor

EXISTING POWER FACTOR %	CORRECTED POWER FACTOR					
	100%	95%	90%	85%	80%	75%
50	1.732	1.403	1.247	1.112	0.982	0.850
52	1.643	1.314	1.158	1.023	0.893	0.761
54	1.558	1.229	1.073	0.938	0.808	0.676
55	1.518	1.189	1.033	0.898	0.768	0.636
56	1.479	1.150	0.994	0.859	0.729	0.597
58	1.404	1.075	0.919	0.784	0.654	0.522
60	1.333	1.004	0.848	0.713	0.583	0.451
62	1.265	0.936	0.780	0.645	0.515	0.383
64	1.201	0.872	0.716	0.581	0.451	0.319
65	1.168	0.839	0.683	0.548	0.418	0.286
66	1.139	0.810	0.654	0.519	0.389	0.257
68	1.078	0.749	0.593	0.458	0.328	0.196
70	1.020	0.691	0.535	0.400	0.270	0.138
72	0.964	0.635	0.479	0.344	0.214	0.082
74	0.909	0.580	0.424	0.289	0.159	0.027
75	0.882	0.553	0.397	0.262	0.132	
76	0.855	0.526	0.370	0.235	0.105	
78	0.802	0.473	0.317	0.182	0.052	
80	0.750	0.421	0.265	0.130		
82	0.698	0.369	0.213	0.078		
84	0.646	0.317	0.161			
85	0.620	0.291	0.135			
86	0.594	0.265	0.109			
88	0.540	0.211	0.055			
90	0.485	0.156				
92	0.426	0.097				
94	0.363	0.034				
95	0.329					

TYPICAL PROBLEM: With a load of 500 KW at 70% power factor, it is desired to find the KVA of capacitors required to correct the power factor to 85%.

SOLUTION: From the table, select the multiplying factor 0.400 corresponding to the existing 70%, and the corrected 85% power factor.

0.400 x 500 = 200 KVA of capacitors required.

POWER FACTOR AND EFFICIENCY EXAMPLE

A squirrel cage induction motor is rated 10 horsepower, 208 volt, three phase and has a nameplate rating of 27.79 amperes. A wattmeter reading indicates 8 kilowatts of consumed (true) power. Calculate apparent power (KVA), power factor, efficiency, internal losses, and size of the capacitor in kilovolts reactive (KVAR) needed to correct the power factor to unity (100%).

Apparent input power: kilovolt-amperes (KVA)
KVA = (E x I x 1.73) / 1000 = (208 x 27.79 x 1.73) / 1000 = **10 KVA**

Power factor (PF) = ratio of true power (KW) to apparent power (KVA)
Kilowatts / kilovolt-amperes = 8 KW / 10 KVA = 0.8 = **80% Power Factor**
80% of the 10-KVA apparent power input performs work.

Motor output in kilowatts = 10 horsepower x 746 watts = 7460 watts = **7.46 KW**
Efficiency = watts out/watts in = 7.46 KW / 8 KW = 0.9325 = **93.25% efficiency**

Internal losses (heat, friction, hysteresis) = 8 KW - 7.46 KW = **0.54 KW** (540 watts)

Kilovolt-amperes reactive (KVAR) (Power stored in motor magnetic field)

KVAR = $\sqrt{KVA^2 - KW^2} = \sqrt{10\ KVA^2 - 8\ KW^2} = \sqrt{100 - 64} = \sqrt{36}$ = **6 KVAR**

The size capacitor needed to equal the motor's stored reactive power is 6 KVAR. (A capacitor stores reactive power in its electrostatic field.)

The power source must supply the current to perform work and maintain the motor's magnetic field. Before power factor correction, this was 27.79 amperes. The motor magnetizing current after power factor correction is supplied by circulation of current between the motor and the electrostatic field of the capacitor and is no longer supplied by power source after initial start up. The motor feeder current after correction to 100% will equal the amount required by the input watts in this case (8 KW x 1000) / (208 volts x 1.73) = **22.23 amperes.**

- Kilo = 1000. For example: 1000 watts = 1 kilowatt.
- Inductive loads (motors, coils) have lagging currents, and capacitive loads have leading currents.
- Inductance and capacitance have opposite effects in a circuit and can cancel each other.

TO FIND INDUCTANCE

Inductance (L):

Inductance is the production of magnetization of electrification in a body by the proximity of a magnetic field or electric charge, or of the electric current in a conductor by the variation of the magnetic field in its vicinity. Expressed in Henrys.

A. To find the total inductance of coils connected in series:

$$L_T = L_1 + L_2 + L_3 + L_4$$

Determine the total inductance of four coils connected in series. Each coil has an inductance of 4 Henrys.

$$L_T = L_1 + L_2 + L_3 + L_4$$

$$= 4 + 4 + 4 + 4 = 16 \text{ HENRYS}$$

B. To find the total inductance of coils connected in parallel.

$$\frac{1}{L_T} = \frac{1}{L_1} + \frac{1}{L_2} + \frac{1}{L_3} + \frac{1}{L_4}$$

Determine the total inductance of four coils connected in parallel: Each coil has an inductance of 4 Henrys.

$$\frac{1}{L_T} = \frac{1}{L_1} + \frac{1}{L_2} + \frac{1}{L_3} + \frac{1}{L_4}$$

$$\frac{1}{L_T} = \frac{1}{4} + \frac{1}{4} + \frac{1}{4} + \frac{1}{4}$$

$$\frac{1}{L_T} = \frac{4}{4} \text{ OR } L_T \times 4 = 1 \times 4 \text{ OR } L_T = \frac{4}{4} = 1 \text{ HENRY}$$

An induction coil is a device consisting of two concentric coils and an interrupter, which changes a low steady voltage into a high intermittent alternating voltage by electromagnetic induction. Most often used as a spark coil.

TO FIND IMPEDANCE

Impedance (Z):

Impedance is the total opposition to an alternating current presented by a circuit. Expressed in ohms.

A. When *VOLTS and AMPERES* are known:

$$\text{IMPEDANCE} = \frac{\text{VOLTS}}{\text{AMPERES}} \quad \text{OR} \quad Z = \frac{E}{I}$$

Determine the impedance of a 120-volt ac circuit that draws a current of 4 amperes.

$$Z = \frac{E}{I} = \frac{120}{4} = 30 \text{ OHMS}$$

B. When *RESISTANCE and REACTANCE* are known:

$$Z = \sqrt{\text{RESISTANCE}^2 + \text{REACTANCE}^2} = \sqrt{R^2 + X^2}$$

Determine the impedance of an ac circuit when the resistance is 6 ohms, and the reactance is 8 ohms.

$$Z = \sqrt{R^2 + X^2} = \sqrt{36 + 64} = \sqrt{100} = 10 \text{ OHMS}$$

C. When *RESISTANCE, INDUCTIVE REACTANCE, and CAPACITIVE REACTANCE* are known:

$$Z = \sqrt{R^2 + (X_L - X_C)^2}$$

Determine the impedance of an ac circuit that has a resistance of 6 ohms, an inductive reactance of 18 ohms, and a capacitive reactance of 10 ohms.

$$Z = \sqrt{R^2 + (X_L - X_C)^2}$$

$$= \sqrt{6^2 + (18 - 10)^2} = \sqrt{6^2 + (8)^2}$$

$$= \sqrt{36 + 64} = \sqrt{100} = 10 \text{ OHMS}$$

TO FIND REACTANCE

Reactance (X):

Reactance in a circuit is the opposition to an alternating current caused by inductance and capacitance, equal to the difference between capacitive and inductive reactance. Expressed in ohms.

A. INDUCTIVE REACTANCE X_L

Inductive reactance is that element of reactance in a circuit caused by self-inductance.

$$X_L = 2 \times 3.1416 \times \text{FREQUENCY} \times \text{INDUCTANCE}$$
$$= 6.28 \times F \times L$$

Determine the reactance of a 4-Henry coil on a 60-cycle ac circuit.

$$X_L = 6.28 \times F \times L = 6.28 \times 60 \times 4 = 1507 \text{ OHMS}$$

B. CAPACITIVE REACTANCE X_C

Capacitive reactance is that element of reactance in a circuit caused by capacitance.

$$X_C = \frac{1}{2 \times 3.1416 \times \text{FREQUENCY} \times \text{CAPACITANCE}}$$

$$= \frac{1}{6.28 \times F \times C}$$

Determine the reactance of a 4 microfarad condenser on a 60-cycle ac circuit.

$$X_C = \frac{1}{6.28 \times F \times C} = \frac{1}{6.28 \times 60 \times 0.000004}$$

$$= \frac{1}{0.0015072} = 663 \text{ OHMS}$$

A Henry is a unit of inductance equal to the inductance of a circuit in which the variation of a current at the rate of 1 ampere per second induces an electromotive force of 1 volt.

FULL-LOAD CURRENT IN AMPERES: DIRECT-CURRENT MOTORS

	Armature Voltage Rating*					
HP	90 V	120 V	180 V	240 V	500 V	550 V
¼	4.0	3.1	2.0	1.6	–	–
⅓	5.2	4.1	2.6	2.0	–	–
½	6.8	5.4	3.4	2.7	–	–
¾	9.6	7.6	4.8	3.8	–	–
1	12.2	9.5	6.1	4.7	–	–
1½	–	13.2	8.3	6.6	–	–
2	–	17	10.8	8.5	–	–
3	–	25	16	12.2	–	–
5	–	40	27	20	–	–
7½	–	58	–	29	13.6	12.2
10	–	76	–	38	18	16
15	–	–	–	55	27	24
20	–	–	–	72	34	31
25	–	–	–	89	43	38
30	–	–	–	106	51	46
40	–	–	–	140	67	61
50	–	–	–	173	83	75
60	–	–	–	206	99	90
75	–	–	–	255	123	111
100	–	–	–	341	164	148
125	–	–	–	425	205	185
150	–	–	–	506	246	222
200	–	–	–	675	330	294

These values of full-load currents* are for motors running at base speed.

*These are average dc quantities.

DIRECT-CURRENT MOTORS

Terminal Markings

Terminal markings are used to tag terminals to which connections are to be made from outside circuits.

Facing the end opposite the drive (commutator end), the standard direction of shaft rotation is counter-clockwise.

A_1 and A_2 indicate armature leads.
S_1 and S_2 indicate series-field leads.
F_1 and F_2 indicate shunt-field leads.

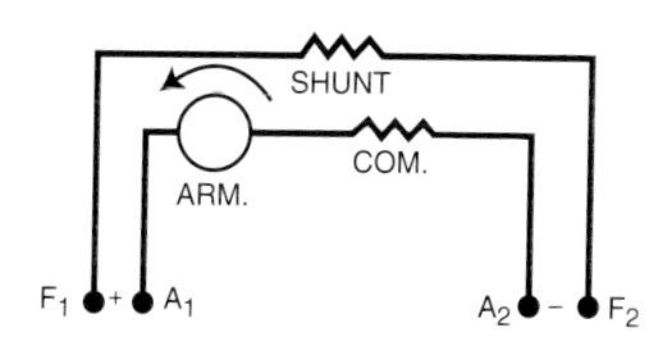

Shunt-Wound Motors

To change rotation, reverse either armature leads or shunt leads. Do not reverse both armature and shunt leads.

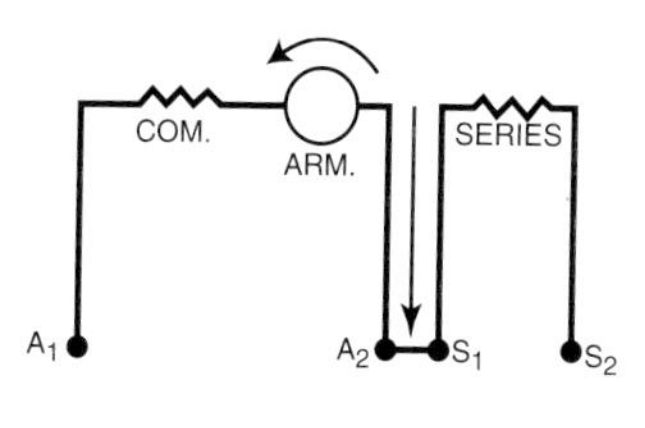

Series-Wound Motors

To change rotation, reverse either armature leads or series leads. Do not reverse both armature and series leads.

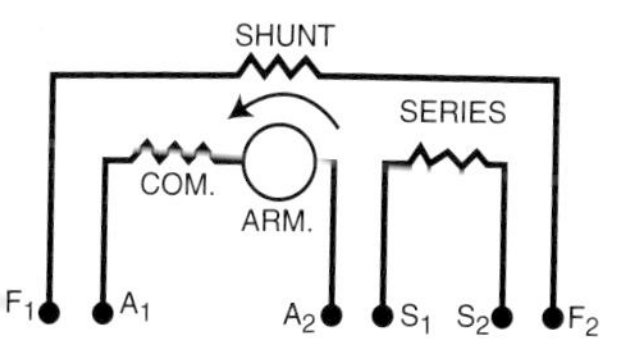

Compound-Wound Motors

To change rotation, reverse either armature leads or both the series and shunt leads. Do not reverse all three sets of leads.

NOTE: Standard rotation for dc generator is clockwise.

FULL-LOAD CURRENT IN AMPERES: SINGLE-PHASE ALTERNATING-CURRENT MOTORS

HP	115 V	200 V	208 V	230 V
1/6	4.4	2.5	2.4	2.2
1/4	5.8	3.3	3.2	2.9
1/3	7.2	4.1	4.0	3.6
1/2	9.8	5.6	5.4	4.9
3/4	13.8	7.9	7.6	6.9
1	16	9.2	8.8	8.0
1 1/2	20	11.5	11	10
2	24	13.8	13.2	12
3	34	19.6	18.7	17
5	56	32.2	30.8	28
7 1/2	80	46	44	40
10	100	57.5	55	50

The voltages listed are rated motor voltages. The currents listed shall be permitted for system voltage ranges of 110 to 120 and 220 to 240 volts.

Source: NFPA 70, *National Electrical Code*®, NFPA, Quincy, MA, 2013, Table 430.248.

SINGLE-PHASE MOTOR USING STANDARD THREE-PHASE STARTER

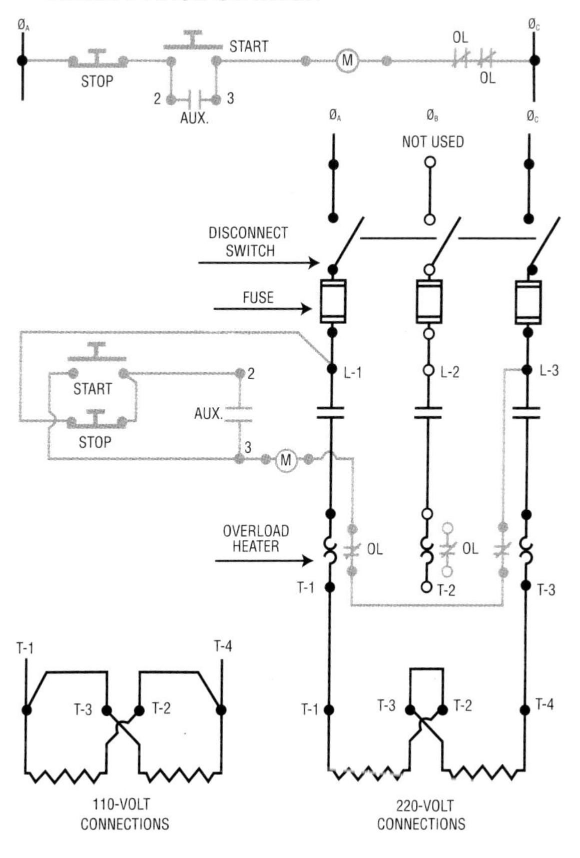

(M) = MOTOR STARTER COIL

SINGLE-PHASE MOTORS

Split-Phase, Squirrel-Cage, Dual-Voltage Motor

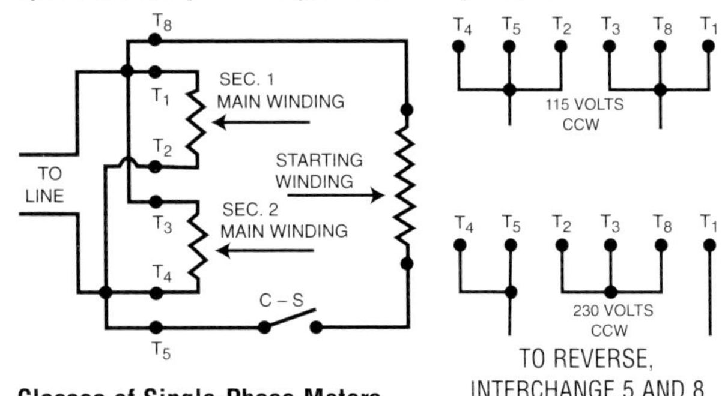

Classes of Single-Phase Motors

1. SPLIT-PHASE
   A. CAPACITOR START
   B. REPULSION START
   C. RESISTANCE START
   D. SPLIT CAPACITOR

2. COMMUTATOR
   A. REPULSION
   B. SERIES

Terminal Color Marking

T_1 BLUE	T_3 ORANGE	T_5 BLACK
T_2 WHITE	T_4 YELLOW	T_8 RED

NOTE: Split-phase motors are usually fractional horsepower. The majority of electric motors used in washing machines, refrigerators, etc. are of the split-phase type.

To change the speed of a split-phase motor, the number of poles must be changed.

1. Addition of running winding
2. Two starting windings and two running windings
3. Consequent pole connections

SINGLE-PHASE MOTORS

Split-Phase, Squirrel-Cage Motor

A. RESISTANCE START:

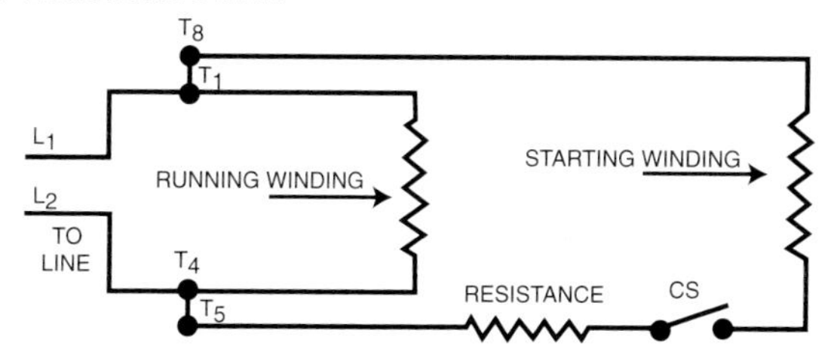

Centrifugal switch (CS) opens after reaching 75% of normal speed.

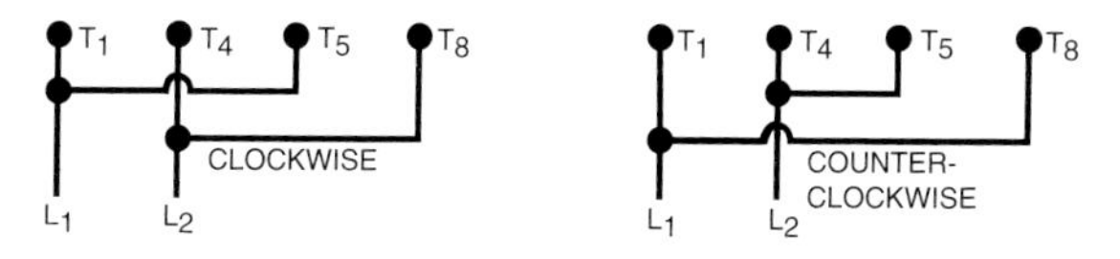

B. CAPACITOR START:

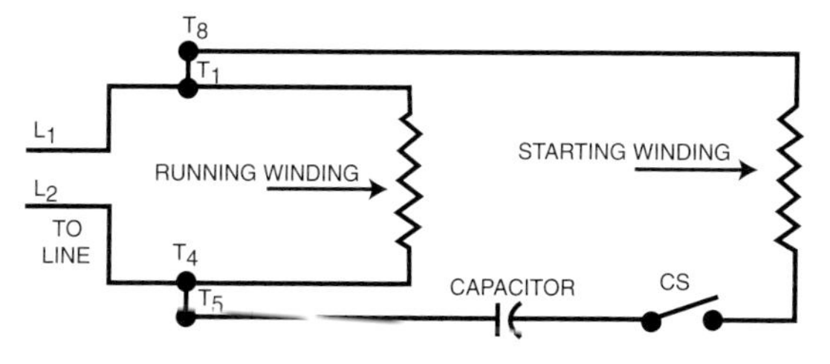

NOTE: 1. A resistance start motor has a resistance connected in series with the starting winding.

2. The capacitor start motor is employed where a high starting torque is required.

RUNNING OVERLOAD UNITS

KIND OF MOTOR	SUPPLY SYSTEM	NUMBER AND LOCATION OF OVERLOAD UNITS SUCH AS TRIP COILS OR RELAYS
1-phase ac or dc	2-wire, 1-phase ac or dc, ungrounded	1 in either conductor
1-phase ac or dc	2-wire, 1-phase ac or dc, one conductor ungrounded	1 in ungrounded conductor
1-phase ac or dc	3-wire, 1-phase ac or dc, grounded neutral conductor	1 in either ungrounded conductor
1-phase ac	Any 3-phase	1 in ungrounded conductor
2-phase ac	3-wire, 2-phase ac, ungrounded	2, one in each phase
2-phase ac	3-wire, 2-phase ac, one conductor grounded	2 in ungrounded conductors
2-phase ac	4-wire, 2-phase ac, grounded or ungrounded	2, one for each phase in ungrounded conductors
2-phase ac	Grounded neutral or 5-wire, 2-phase ac, ungrounded	2, one for each phase in any ungrounded phase wire
3-phase ac	Any 3-phase	3, one in each phase*

*Exception: An overload unit in each phase shall not be required where overload protection is provided by other approved means.
Source: NFPA 70, *National Electrical Code®*, NFPA, Quincy, MA, 2013, Table 430.37.

MOTOR BRANCH-CIRCUIT PROTECTIVE DEVICES: MAXIMUM RATING OR SETTING

	Percent of Full-Load Current			
Type of Motor	Nontime Delay Fuse[1]	Dual Element (Time-Delay) Fuse[1]	Instantaneous Trip Breaker	Inverse Time Breaker[2]
Single-phase motors	300	175	800	250
AC polyphase motors other than wound rotor	300	175	800	250
Squirrel Cage				
–other than Design B energy-efficient	300	175	800	250
Design B energy-efficient	300	175	1100	250
Synchronous[3]	300	175	800	250
Wound rotor	150	150	800	150
Direct current (constant voltage)	150	150	250	150

For certain exceptions to the values specified, see 430.54.
1 The values in the Nontime Delay Fuse column apply to Time-Delay Class CC fuses.
2 The values given in the last column also cover the ratings of nonadjustable inverse time types of circuit breakers that may be modified as in 430.52. (C)(1), Exception No. 1 and No. 2.
3 Synchronous motors of the low-torque, low-speed type (usually 450 rpm or lower), such as are used to drive reciprocating compressors, pumps, and so forth, that start unloaded, do not require a fuse rating or circuit-breaker setting in excess of 200% of full-load current.
Source: NFPA 70, *National Electrical Code®*, NFPA, Quincy, MA, 2013, Table 430.52.

FULL-LOAD CURRENT: THREE-PHASE ALTERNATING-CURRENT MOTORS

	Induction-Type Squirrel Cage and Wound Rotor (Amperes)							Synchronous-Type Unity Power Factor* (Amperes)			
HP	115 Volts	200 Volts	208 Volts	230 Volts	460 Volts	575 Volts	2300 Volts	230 Volts	460 Volts	575 Volts	2300 Volts
½	4.4	2.5	2.4	2.2	1.1	0.9	-	-	-	-	-
¾	6.4	3.7	3.5	3.2	1.6	1.3	-	-	-	-	-
1	8.4	4.8	4.6	4.2	2.1	1.7	-	-	-	-	-
1½	12.0	6.9	6.6	6.0	3.0	2.4	-	-	-	-	-
2	13.6	7.8	7.5	6.8	3.4	2.7	-	-	-	-	-
3	-	11.0	10.6	9.6	4.8	3.9	-	-	-	-	-
5	-	17.5	16.7	15.2	7.6	6.1	-	-	-	-	-
7½	-	25.3	24.2	22	11	9	-	-	-	-	-
10	-	32.2	30.8	28	14	11	-	-	-	-	-
15	-	48.3	46.2	42	21	17	-	-	-	-	-
20	-	62.1	59.4	54	27	22	-	-	-	-	-
25	-	78.2	74.8	68	34	27	-	53	26	21	-
30	-	92	88	80	40	32	-	63	32	26	-
40	-	120	114	104	52	41	-	83	41	33	-
50	-	150	143	130	65	52	-	104	52	42	-
60	-	177	169	154	77	62	16	123	61	49	12
75	-	221	211	192	96	77	20	155	78	62	15
100	-	285	273	248	124	99	26	202	101	81	20
125	-	359	343	312	156	125	31	253	126	101	25
150	-	414	396	360	180	144	37	302	151	121	30
200	-	552	528	480	240	192	49	400	201	161	40
250	-	-	-	-	302	242	60	-	-	-	-
300	-	-	-	-	361	289	72	-	-	-	-
350	-	-	-	-	414	336	83	-	-	-	-
400	-	-	-	-	477	382	95	-	-	-	-
450	-	-	-	-	515	412	103	-	-	-	-
500	-	-	-	-	590	472	110		-	-	-

The voltages listed are rated motor voltages. The currents listed shall be permitted for system voltage ranges of 110 to 120, 220 to 240, 440 to 480, and 550 to 1000 volts.

*For 90% and 80% power factor, the figures shall be multiplied by 1.1 and 1.25, respectively.

Source: NFPA 70, *National Electrical Code*®, NFPA, Quincy, MA, 2013, Table 430.250.

FULL-LOAD CURRENT AND OTHER DATA: THREE-PHASE AC MOTORS

MOTOR HORSEPOWER		MOTOR AMPERE	SIZE BREAKER *	SIZE STARTER	HEATER AMPERE **	SIZE WIRE	SIZE CONDUIT ***
½	230 V	2.2	15	00	2.530	12	¾"
	460	1.1	15	00	1.265	12	¾
¾	230	3.2	15	00	3.680	12	¾
	460	1.6	15	00	1.840	12	¾
1	230	4.2	15	00	4.830	12	¾
	460	2.1	15	00	2.415	12	¾
1½	230	6.0	15	00	6.900	12	¾
	460	3.0	15	00	3.450	12	¾
2	230	6.8	15	0	7.820	12	¾
	460	3.4	15	00	3.910	12	¾
3	230	9.6	20	0	11.040	12	¾
	460	4.8	15	0	5.520	12	¾
5	230	15.2	30	1	17.480	12	¾
	460	7.6	15	0	8.740	12	¾
7½	230	22	45	1	25.300	10	¾
	460	11	20	1	12.650	12	¾
10	230	28	60	2	32.200	10	¾
	460	14	30	1	16.100	12	¾
15	230	42	70	2	48.300	6	1
	460	21	40	2	24.150	10	¾
20	230	54	100	3	62.100	4	1
	460	27	50	2	31.050	10	¾
25	230	68	100	3	78.200	4	1½
	460	34	50	2	39.100	8	1
30	230	80	125	3	92.000	3	1½
	460	40	70	3	46.000	8	1
40	230	104	175	4	119.600	1	1½
	460	52	100	3	59.800	6	1
50	230	130	200	4	149.500	00	2
	460	65	150	3	74.750	4	1½
60	230	154	250	5	177.10	000	2
	460	77	200	4	88.55	3	1½

FULL-LOAD CURRENT AND OTHER DATA: THREE-PHASE, AC MOTORS

MOTOR HORSEPOWER		MOTOR AMPERE	SIZE BREAKER *	SIZE STARTER	HEATER AMPERE **	SIZE WIRE	SIZE CONDUIT ***
75	230 V	192	300	5	220.80	250 kcmil	2½"
	460	96	200	4	110.40	1	1½
100	230	248	400	5	285.20	350 kcmil	3
	460	124	200	4	142.60	2/0	2
125	230	312	500	6	358.80	600 kcmil	3½
	460	156	250	5	179.40	000	2
150	230	360	600	6	414.00	700 kcmil	4
	460	180	300	5	207.00	0000	2½

* Overcurrent device may have to be increased due to starting current and load conditions. See *NEC* 430.52, Table 430.52. Wire size based on 75°C (167°F) terminations and 75°C (167°F) insulation.

** Overload heater must be based on motor nameplate and sized per *NEC* 430.32.

*** Conduit size based on rigid metal conduit with some spare capacity. For minimum size and other conduit types, see *NEC* Appendix C, or *Ugly's* pages 83–103.

MOTOR AND MOTOR CIRCUIT CONDUCTOR PROTECTION

Motors can have large starting currents three to five times or more than that of the actual motor current. In order for motors to start, the motor and motor circuit conductors are allowed to be protected by circuit breakers and fuses at values that are higher than the actual motor and conductor ampere ratings. These larger overcurrent devices do not provide overload protection and will only open upon short circuits or ground faults. Overload protection must be used to protect the motor based on the actual nameplate amperes of the motor. This protection is usually in the form of heating elements in manual or magnetic motor starters. Small motors such as waste disposal motors have a red overload reset button built into the motor.

General Motor Rules

- Use full-load current from Tables instead of nameplate.
- Branch Circuit Conductors: Use 125% of full-load current to find conductor size.
- Branch Circuit OCP Size: Use percentages given in Tables for full-load current. (*Ugly's* page 32)
- Feeder Conductor Size: 125% of largest motor and sum of the rest.
- Feeder OCP: Use largest OCP plus rest of full-load currents.

(See examples on Ugly's *page 42.)*

MOTOR BRANCH CIRCUIT AND FEEDER EXAMPLE

General Motor Applications

Branch-Circuit Conductors:

Use full-load, three-phase currents, from the table on *Ugly's* page 39 or 2014 *NEC* Table 430.250, 50-HP, 480-volt, 3-phase, motor design B, 75-degree terminations = 65 amperes
125% of full-load current [*NEC* 430.22(A)] (*Ugly's* page 41)
125% of 65 A = **81.25 amperes** conductor selection ampacity

Branch Circuit Overcurrent Device: *NEC* 430.52 (C)(1)

(Branch circuit short-circuit and ground fault protection)
Use percentages given in *Ugly's* page 38 or 2014 *NEC* 430.52 for **Type** of circuit breaker or fuse used.
50-HP, 480-V, 3-Ph motor = 65 amperes (*Ugly's* page 39).
Nontime fuse = 300% (*Ugly's* page 38).
300% of 65A = 195 A. *NEC* 430.52(C1)(EX1) Next size allowed *NEC* 240.6A = **200-ampere fuse**.

Feeder Connectors:

For 50-HP and 30-HP, 480-volt, 3-phase, design B motors on same feeder:
Use 125% of the largest full-load current and 100% of the rest. (*NEC* 430.24)
50-HP, 480-V, 3-Ph motor = 65 A; 30-HP, 480-V, 3-Ph motor = 40 A
(125% of 65 A) + 40 A = **121.25 A** conductor selection ampacity

Feeder Overcurrent Device: *NEC* 430.62(A) (specific load)

(Feeder short-circuit and ground-fault protection)
Use largest overcurrent protection device plus full-load currents of the rest of the motors.
50 HP = 200 A fuse (65 FLC)
30 HP = 125 A fuse (40 FLC)
200 A fuse + 40 A (FLC) = 240 A. Do not exceed this value on feeder.
Go down to a **225-A** fuse.

LOCKED-ROTOR CODE LETTERS

Code Letter	Kilovolt-Ampere/ Horsepower with Locked Rotor	Code Letter	Kilovolt-Ampere/ Horsepower with Locked Rotor
A	0–3.14	L	9.0–9.99
B	3.15–3.54	M	10.0–11.19
C	3.55–3.99	N	11.2–12.49
D	4.0–4.49	P	12.5–13.99
E	4.5–4.99	R	14.0–15.99
F	5.0–5.59	S	16.0–17.99
G	5.6–6.29	T	18.0–19.99
H	6.3–7.09	U	20.0–22.39
J	7.1–7.99	V	22.4 and up
K	8.0–8.99		

Source: NFPA 70, *National Electrical Code®*, NFPA, Quincy, MA, 2013, Table 430.7(B), as modified.

The *National Electrical Code®* requires that all alternating-current motors rated ½ horsepower or more (except for polyphase wound-rotor motors) must have code letters on their nameplates indicating motor input with locked rotor (in kilovolt-amperes per horsepower). If you know the horsepower and voltage rating of a motor and its "Locked KVA/Horsepower" (from the above table), you can calculate the locked-rotor current using the following formulas.

Single-Phase Motors:

$$\text{Locked-Rotor Current} = \frac{HP \times KVA_{hp} \times 1000}{E}$$

Three-Phase Motors:

$$\text{Locked-Rotor Current} = \frac{HP \times KVA_{hp} \times 1000}{E \times 1.73}$$

Example: What is the maximum locked-rotor current for a 480-volt, 25-horsepower, code letter F motor?
(From the above table, code letter F = 5.59 KVA_{hp})

$$I = \frac{HP \times KVA_{hp} \times 1000}{E \times 1.73} = \frac{25 \times 5.59 \times 1000}{480 \times 1.73} = \textbf{168.29 amperes}$$

MAXIMUM MOTOR LOCKED-ROTOR CURRENT IN AMPERES, SINGLE PHASE

HP	115 V	208 V	230 V	HP	115 V	208 V	230 V
½	58.8	32.5	29.4	3	204	113	102
¾	82.8	45.8	41.4	5	336	186	168
1	96	53	48	7½	480	265	240
1½	120	66	60	10	600	332	300
2	144	80	72				

Note: For use only with 430.110, 440.12, 440.41, and 455.8(C).
Source: NFPA 70, *National Electrical Code*®, NFPA, Quincy, MA, 2013, Table 430.251(A), as modified.

MAXIMUM MOTOR LOCKED-ROTOR CURRENT IN AMPERES, TWO AND THREE PHASE, DESIGN B, C, AND D*

HP	115 V	200 V	208 V	230 V	460 V	575 V
½	40	23	22.1	20	10	8
¾	50	28.8	27.6	25	12.5	10
1	60	34.5	33	30	15	12
1½	80	46	44	40	20	16
2	100	57.5	55	50	25	20
3	–	73.6	71	64	32	25.6
5	–	105.8	102	92	46	36.8
7½	–	146	140	127	63.5	50.8
10	–	186.3	179	162	81	64.8
15	–	267	257	232	116	93
20	–	334	321	290	145	116
25	–	420	404	365	183	146
30	–	500	481	435	218	174
40	–	667	641	580	290	232
50	–	834	802	725	363	290
60	–	1001	962	870	435	348
75	–	1248	1200	1085	543	434
100	–	1668	1603	1450	725	580
125	–	2087	2007	1815	908	726
150	–	2496	2400	2170	1085	868
200	–	3335	3207	2900	1450	1160

* Design A motors are not limited to a maximum starting current or locked-rotor current.
Note: For use only with 430.110, 440.12, 440.41, and 455.8(C).
Source: NFPA 70, *National Electrical Code*®, NFPA, Quincy, MA, 2013, Table 430.251(B), as modified.

THREE-PHASE AC MOTOR WINDINGS AND CONNECTIONS

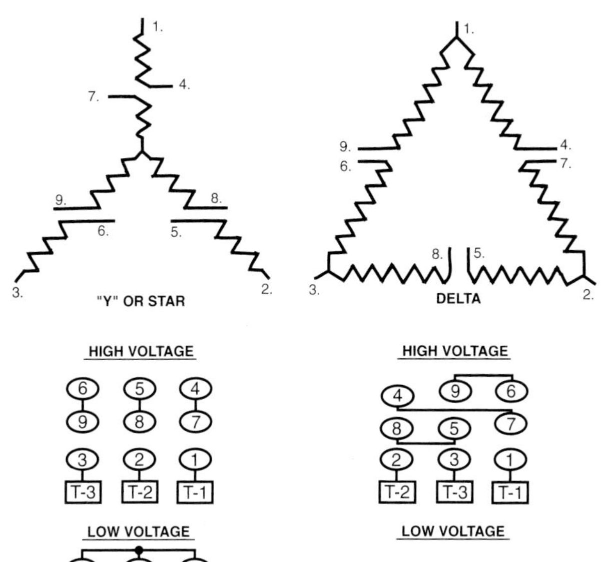

LOW VOLTAGE

6 5 4
9 8 7
3 2 1
T-3 T-2 T-1

LOW VOLTAGE

4 9 6
8 5 7
2 3 1
T-2 T-3 T-1

NOTE: 1. The most important part of any motor is the nameplate. Check the data given on the plate before making the connections.

2. To change rotation direction of 3-phase motor, swap any two T-leads.

THREE-WIRE STOP-START STATION

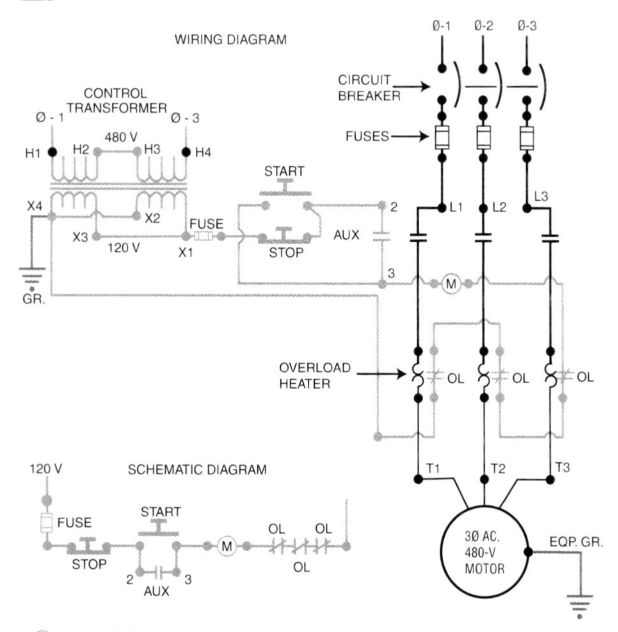

(M) = MOTOR STARTER COIL

NOTE: CONTROLS AND MOTOR ARE OF DIFFERENT VOLTAGES.

TWO THREE-WIRE STOP-START STATIONS

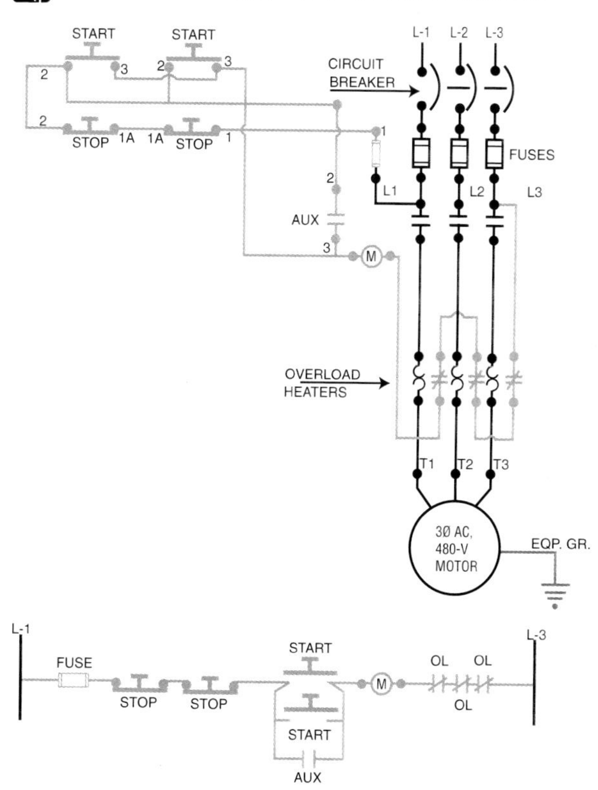

(M) = MOTOR STARTER COIL

NOTE: Controls and motor are of the same voltage.
If low-voltage controls are used, see *Ugly's* page 46 for control transformer connections.

HAND-OFF AUTOMATIC CONTROL

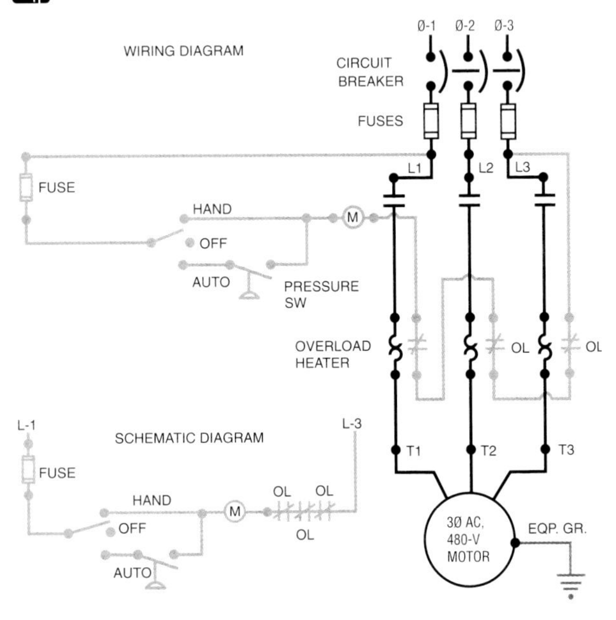

NOTE: Controls and motor are of the same voltage.
If low-voltage controls are used, see *Ugly's* page 46 for control transformer connections.

JOGGING WITH CONTROL RELAY

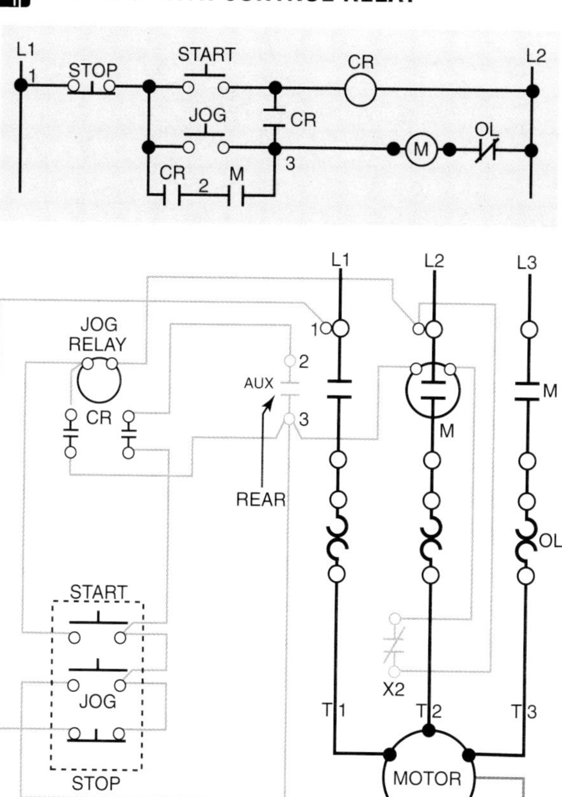

Jogging circuits are used when machines must be operated momentarily for inching (as in set-up or maintenance). The jog circuit allows the starter to be energized only as long as the jog button is depressed.

VOLTAGE DROP CALCULATIONS: INDUCTANCE NEGLIGIBLE

Vd = Voltage Drop
I = Current in Conductor (Amperes)
L = One-way Length of Circuit (Feet)
Cm = Cross Section Area of Conductor (Circular Mils) (page 71)
K = Resistance in ohms of 1 circular mil foot of conductor
K = 12.9 for Copper Conductors @ 75°C (167°F)
K = 21.2 for Aluminum Conductors @ 75°C (167°F)

NOTE: K value changes with temperature. See *NEC* Chapter 9, Table 8, Notes.

Single-Phase Circuits

$$Vd = \frac{2K \times L \times I}{Cm} \quad \text{or} \quad {}^{*}Cm = \frac{2K \times L \times I}{Vd}$$

Three-Phase Circuits

$$Vd = \frac{1.73K \times L \times I}{Cm} \quad \text{or} \quad {}^{*}Cm = \frac{1.73K \times L \times I}{Vd}$$

*NOTE: Always check ampacity tables to ensure conductor's ampacity is equal to load after voltage drop calculation.

Refer to *Ugly's* pages 71–81 for conductor size, type, and ampacity.
See *Ugly's* pages 51–52 for examples.

VOLTAGE DROP EXAMPLES

Distance (One-Way) for 2% Voltage Drop for 120 or 240 Volts, Single Phase (60°C [140°F] Insulation and Terminals)

AMPS	VOLTS	12 AWG	10 AWG	8 AWG	6 AWG	4 AWG	3 AWG	2 AWG	1 AWG	1/0 AWG
20	120	30	48	77	122	194	245	309	389	491
	240	60	96	154	244	388	490	618	778	982
30	120		32	51	81	129	163	206	260	327
	240		64	102	162	258	326	412	520	654
40	120			38	61	97	122	154	195	246
	240			76	122	194	244	308	390	492
50	120				49	78	98	123	156	196
	240				98	156	196	246	312	392
60	120					65	82	103	130	164
60	240					130	164	206	260	328
70	240					111	140	176	222	281
80	240						122	154	195	246
90	240							137	173	218
100	240								156	196

(SEE FOOTNOTES ON PAGE 51 CONCERNING CIRCUIT LOAD LIMITATIONS.)

VOLTAGE DROP EXAMPLES

Typical Voltage Drop Values Based on Conductor Size and One-Way Length* (60°C [140°F] Termination and Insulation)

	25 FEET							
	12 AWG	10 AWG	8 AWG	6 AWG	4 AWG	3 AWG	2 AWG	1 AWG
20 A	1.98	1.24	0.78	0.49	0.31	0.25	0.19	0.15
30		1.86	1.17	0.74	0.46	0.37	0.29	0.23
40			1.56	0.98	0.62	0.49	0.39	0.31
50				1.23	0.77	0.61	0.49	0.39
60					0.93	0.74	0.58	0.46

	50 FEET							
	12 AWG	10 AWG	8 AWG	6 AWG	4 AWG	3 AWG	2 AWG	1 AWG
20 A	3.95	2.49	1.56	0.98	0.62	0.49	0.39	0.31
30		3.73	2.34	1.47	0.93	0.74	0.58	0.46
40			3.13	1.97	1.24	0.98	0.78	0.62
50				2.46	1.55	1.23	0.97	0.77
60					1.85	1.47	1.17	0.92

	75 FEET							
	12 AWG	10 AWG	8 AWG	6 AWG	4 AWG	3 AWG	2 AWG	1 AWG
20 A	5.93	3.73	2.34	1.47	0.93	0.74	0.58	0.46
30		5.59	3.52	2.21	1.39	1.10	0.87	0.69
40			4.69	2.95	1.85	1.47	1.17	0.92
50				3.69	2.32	1.84	1.46	1.16
60					2.78	2.21	1.75	1.39

	100 FEET							
	12 AWG	10 AWG	8 AWG	6 AWG	4 AWG	3 AWG	2 AWG	1 AWG
20 A	7.90	4.97	3.13	1.97	1.24	0.98	0.78	0.62
30		7.46	4.69	2.95	1.85	1.47	1.17	0.92
40			6.25	3.93	2.47	1.96	1.56	1.23
50				4.92	3.09	2.45	1.94	1.54
60					3.71	2.94	2.33	1.85

	125 FEET							
	12 AWG	10 AWG	8 AWG	6 AWG	4 AWG	3 AWG	2 AWG	1 AWG
20 A	9.88	6.21	3.91	2.46	1.55	1.23	0.97	0.77
30		9.32	5.86	3.69	2.32	1.84	1.46	1.16
40			7.81	4.92	3.09	2.45	1.94	1.54
50				6.15	3.86	3.06	2.43	1.93
60					4.64	3.68	2.92	2.31

	150 FEET							
	12 AWG	10 AWG	8 AWG	6 AWG	4 AWG	3 AWG	2 AWG	1 AWG
20 A	11.85	7.46	4.69	2.95	1.85	1.47	1.17	0.92
30		11.18	7.03	4.42	2.78	2.21	1.75	1.39
40			9.38	5.90	3.71	2.94	2.33	1.85
50				7.37	4.64	3.68	2.92	2.31
60					5.56	4.41	3.50	2.77

A 2-wire, 20-ampere circuit using 12 AWG with a one-way distance of 25 feet will drop 1.98 volts;
120 volts - 1.98 volts = 118.02 volts as the load voltage.
240 volts - 1.98 volts = 238.02 volts as the load voltage.

*Better economy and efficiency will result using the voltage drop method on page 50.
A continuous load cannot exceed 80% of the circuit rating.
A motor or heating load cannot exceed 80% of the circuit rating.
For motor overcurrent devices and conductor sizing, see *Ugly's* pages 41–42.

VOLTAGE DROP CALCULATION EXAMPLES

Single-Phase Voltage Drop

What is the voltage drop of a 240-volt, single-phase circuit consisting of #8 THWN copper conductors feeding a 30-ampere load that is 150 feet in length?

Voltage Drop Formula (see *Ugly's* page 50)

$$\mathbf{Vd} = \frac{\mathbf{2K \times L \times I}}{\mathbf{Cm}} = \frac{2 \times 12.9 \times 150 \times 30}{16{,}510} = \frac{116{,}100}{16{,}510} = \mathbf{7} \text{ volts}$$

Percentage voltage drop = 7 volts/240 volts = 0.029 = **2.9%**
Voltage at load = 240 volts - 7 volts = **233 volts**

Three-Phase Voltage Drop

What is the voltage drop of a 480-volt, 3-phase circuit consisting of 250-kcmil THWN copper conductors that supply a 250-ampere load that is 500 feet from the source?

Voltage Drop Formula (see *Ugly's* page 50)
250 kcmil = 250,000 circular mils

$$\mathbf{Vd} = \frac{\mathbf{1.73K \times L \times I}}{\mathbf{Cm}} = \frac{1.73 \times 12.9 \times 500 \times 250}{250{,}000} = \frac{2{,}789{,}625}{250{,}000} = \mathbf{11} \text{ volts}$$

Percentage voltage drop = 11 volts/480 volts = .0229 = **2.29%**
Voltage at load = 480 volts - 11 volts = **469** volts

NOTE: Always check ampacity tables for conductors selected.

Refer to *Ugly's* pages 71–78 for conductor size, type, and ampacity.

SHORT-CIRCUIT CALCULATION

(Courtesy of Cooper Bussmann)

Basic Short-Circuit Calculation Procedure

1. Determine transformer full-load amperes from either:
   a) Nameplate
   b) Formula:

3ø transf. $$I_{l.l.} = \frac{KVA \times 1000}{E_{L-L} \times 1.732}$$

1ø transf. $$I_{l.l.} = \frac{KVA \times 1000}{E_{L-L}}$$

2. Find transformer multiplier.

$$\text{Multiplier} = \frac{100}{^{*}\%Z_{trans}}$$

3. Determine transformer let-thru short-circuit current.**

$$I_{S.C.} = I_{l.l.} \times \text{Multiplier}$$

4. Calculate "f" factor.

3Ø faults $$f = \frac{1.732 \times L \times I_{3Ø}}{C \times E_{L-L}}$$

1Ø line-to-line (L-L) faults on 1Ø Center Tapped Transformer $$f = \frac{2 \times L \times I_{L-L}}{C \times E_{L-L}}$$

1Ø line-to-neutral (L-N) faults on 1Ø Center Tapped Transformer $$f = \frac{2 \times L \times I_{L-N}^{***}}{C \times E_{L-N}}$$

L = Length (feet) of conductor to the fault
C = Constant from Table C (page 55) for conductors & busway. For parallel runs, multiply C valuoc by tho number of conductors per phase
I = Available short-circuit current in amperes at beginning of circuit.

5. Calculate "M" (multiplier) $$M = \frac{1}{1 + f}$$

6. Calculate the available short-circuit symmetrical RMS current at the point of fault.

$$I_{S.C.\ sym\ RMS} = I_{S.C.} \times M$$

SHORT-CIRCUIT CALCULATION

(Courtesy of Cooper Bussmann)

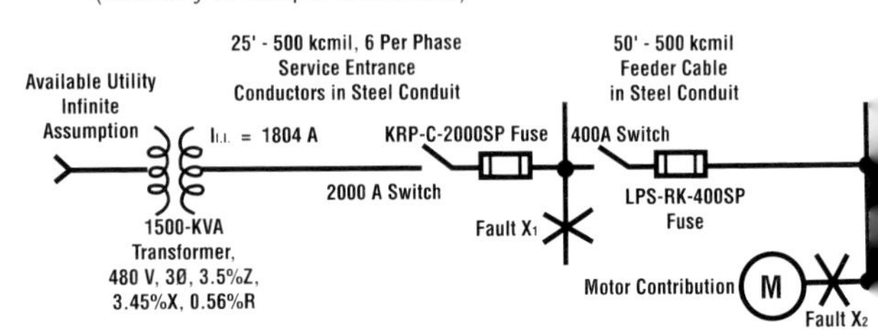

Example: Short-Circuit Calculation

(Fault #1)

1. $$I_{l.l.} = \frac{KVA \times 1000}{E_{L\text{-}L} \times 1.732} = \frac{1500 \times 1000}{480 \times 1.732} = 1804\text{ A}$$

2. $$\text{Multiplier} = \frac{100}{{}^*\%Z_{trans}} = \frac{100}{3.5} = 28.57$$

3. $$I_{S.C.} = 1804 \times 28.57 = 51{,}540\text{ A}$$

4. $$f = \frac{1.732 \times L \times I_{3\emptyset}}{C \times E_{L\text{-}L}} = \frac{1.73 \times 25 \times 51{,}540}{6 \times 22{,}185 \times 480} = 0.0349$$

5. $$M = \frac{1}{1 + f} = \frac{1}{1 + 0.0349} = 0.9663$$

6. $$I_{S.C.\ sym\ RMS} = I_{S.C.} \times M = 51{,}540 \times 0.9663 = 49{,}803\text{ A}$$

   $$I_{S.C.\ motor\ contrib} = 4 \times 1{,}804 = 7{,}216\text{ A}$$

   $$I_{total\ S.C.\ sym\ RMS} = 49{,}803 + 7{,}216 = 57{,}019\text{ A}$$

(Fault #2)

4. Use $I_{S.C.\ sym\ RMS}$ @ Fault X_1 to calculate "f"

$$f = \frac{1.73 \times 50 \times 49{,}803}{22{,}185 \times 480} = 0.4050$$

5. $$M = \frac{1}{1 + 0.4050} = 0.7117$$

6. $$I_{S.C.\ sym\ RMS} = 49{,}803 \times 0.7117 = 35{,}445\text{ A}$$

   $$I_{sym\ motor\ contrib} = 4 \times 1{,}804 = 7{,}216\text{ A}$$

   $$I_{total\ S.C.\ sym\ RMS} = 35{,}445 + 7{,}216 = 42{,}661\text{ A}$$

TABLE "C"

"C" Values for Conductors (Short-Circuit Calculation) (Courtesy of Cooper Bussmann)

AWG or MCM	Copper Three Single Conductors Steel Conduit			Nonmagnetic Conduit			Copper Three Conductor Cable Steel Conduit			Nonmagnetic Conduit		
	600V	5KV	15KV	600V	5KV	15KV	600V	5KV	15KV	600V	5KV	15KV
12	617	617	617	617	617	617	617	617	617	617	617	617
10	981	981	981	981	981	981	981	981	981	981	981	981
8	1557	1551	1557	1558	1555	1558	1559	1557	1559	1559	1558	1559
6	2425	2406	2389	2430	2417	2406	2431	2424	2414	2433	2428	2420
4	3806	3750	3695	3825	3789	3752	3830	3811	3778	3837	3823	3798
3	4760	4760	4760	4802	4802	4802	4760	4790	4760	4802	4802	4802
2	5906	5736	5574	6044	5926	5809	5989	5929	5827	6087	6022	5957
1	7292	7029	6758	7493	7306	7108	7454	7364	7188	7579	7507	7364
1/0	8924	8543	7973	9317	9033	8590	9209	9086	8707	9472	9372	9052
2/0	10755	10061	9389	11423	10877	10318	11244	11045	10500	11703	11528	11052
3/0	12843	11804	11021	13923	13048	12360	13656	13333	12613	14410	14118	13461
4/0	15082	13605	12542	16673	15351	14347	16391	15890	14813	17482	17019	16012
250	16483	14924	13643	18593	17120	15865	18310	17850	16465	19779	19352	18001
300	18176	16292	14768	20867	18975	17408	20617	20051	18318	22524	21938	20163
350	19703	17385	15678	22736	20526	18672	22646	21914	19821	24904	24126	21982
400	20565	18235	16365	24296	21786	19731	24253	23371	21042	26915	26044	23517
500	22185	19172	17492	26706	23277	21329	26980	25449	23125	30028	28712	25916
600	22965	20567	17962	28033	25203	22097	28752	27974	24896	32236	31258	27766
750	24136	21386	18888	28303	25430	22690	31050	30024	26932	32404	31338	28303
1000	25273	22539	19923	31490	28083	24887	33864	32688	29320	37197	35748	31959

(continues)

SHORT-CIRCUIT CALCULATION

(Courtesy of Cooper Bussmann)

NOTES:

*Transformer impedance (Z) helps to determine what the short circuit current will be at the transformer secondary. Transformer impedance is determined as follows:

The transformer secondary is short circuited. Voltage is applied to the primary, which causes full-load current to flow in the secondary. This applied voltage divided by the rated primary voltage is the impedance of the transformer.

Example:

For a 480-volt rated primary, if 9.6 volts causes secondary full-load current to flow through the shorted secondary, the transformer impedance is $9.6 \div 480 = .02 = 2\%Z$.

In addition, U.L. listed transformers 25KVA and larger have a ± 10% impedance tolerance. Short-circuit amperes can be affected by this tolerance.

**Motor short-circuit contribution, if significant, may be added to the transformer secondary short-circuit current value as determined in Step 3. Proceed with this adjusted figure through Steps 4, 5, and 6. A practical estimate of motor short-circuit contribution is to multiply the total motor current in amperes by 4.

***The L-N fault current is higher than the L-L fault current at the secondary terminals of a single-phase center-tapped transformer. The short-circuit current available (I) for this case in Step 4 should be adjusted at the transformer terminals as follows:

At L-N center tapped transformer terminals,

$I_{L\text{-}N} = 1.5 \times I_{L\text{-}L}$ at Transformer Terminals.

COMPONENT PROTECTION

(Courtesy of Cooper Bussmann)

How to Use Current-Limitation Charts

EXAMPLE: An 800-A circuit and an 800-A, low-peak, current-limiting, time-delay fuse

How to Use the Let-Through Charts:

Using the example above, one can determine the pertinent let-through data for the KRP-C-800SP ampere, low-peak fuse. The let-through chart pertaining to the 800-A, low-peak fuse is illustrated.

A. Determine the PEAK let-through CURRENT.

1. Enter the chart on the Prospective Short-Circuit Current scale at 86,000 amperes and proceed vertically until the 800-A fuse curve is intersected.
2. Follow horizontally until the Instantaneous Peak Let-Through Current scale is intersected.
3. Read the PEAK let-through CURRENT as 49,000 A. (If a fuse had not been used, the peak current would have been 198,000 A.)

B. Determine the APPARENT PROSPECTIVE RMS SYMMETRICAL let-through CURRENT.

1. Enter the chart on the Prospective Short-Circuit current scale at 86,000 A and proceed vertically until the 800-A fuse curve is intersected.
2. Follow horizontally until line A-B is intersected.
3. Proceed vertically down to the Prospective Short-Circuit Current.
4. Read the APPARENT PROSPECTIVE RMS SYMMETRICAL let-through CURRENT as 21,000 A. (The RMS SYMMETRICAL let-through CURRENT would be 86,000 A if there were no fuse in the circuit.)

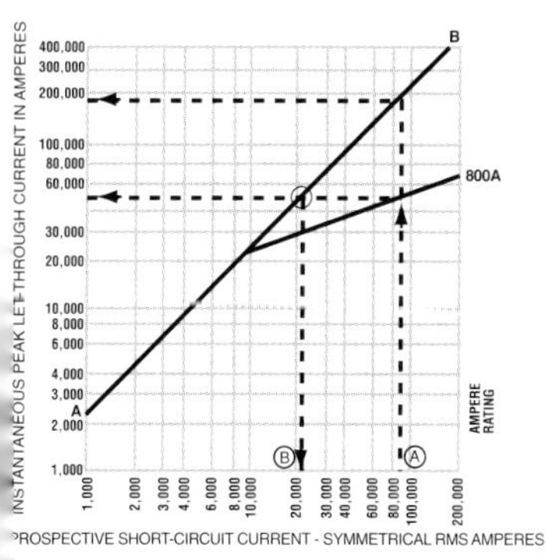

The data that can be obtained from the Fuse Let-Through Charts and their physical effects are:

1) Peak let-through current: Mechanical forces
2) Apparent prospective RMS symmetrical let-through current: Heating effect
3) Clearing time: Less than ½ cycle when fuse is in its current-limiting range (beyond where fuse curve intersects A-B line)

Ⓐ I_{RMS} Available = **86,000 A**
Ⓑ I_{RMS} Let-Through = **21,000 A**
Ⓒ I_P Available = **198,000 A**
Ⓓ I_P Let-Through = **49,000 A**

SINGLE-PHASE TRANSFORMER CONNECTIONS

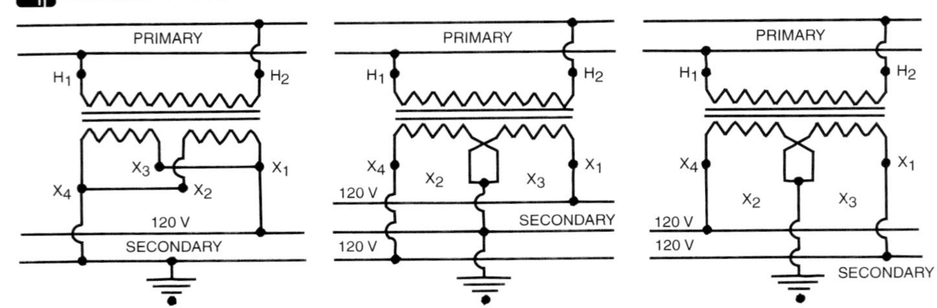

Single phase to supply 120-volt lighting load. Often used for single customer.

Single phase to supply 120/240-V, 3-wire lighting and power load. Used in urban distribution circuits.

Single phase for power. Used for small industrial applications.

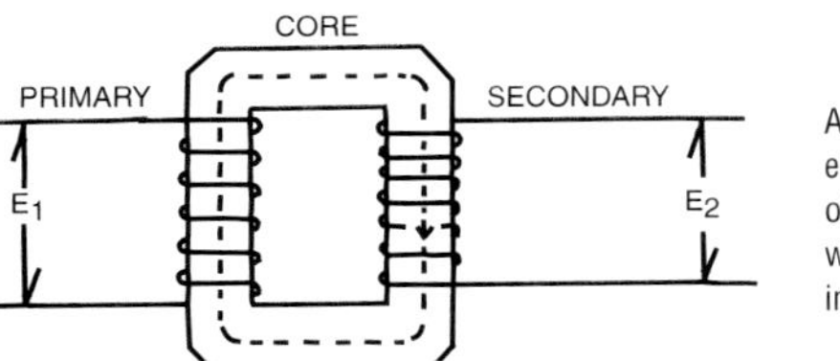

A transformer is a stationary induction device for transferring electrical energy from one circuit to another without change of frequency. A transformer consists of two coils or windings wound upon a magnetic core of soft iron laminations and insulated from one another.

BUCK-AND-BOOST TRANSFORMER CONNECTIONS

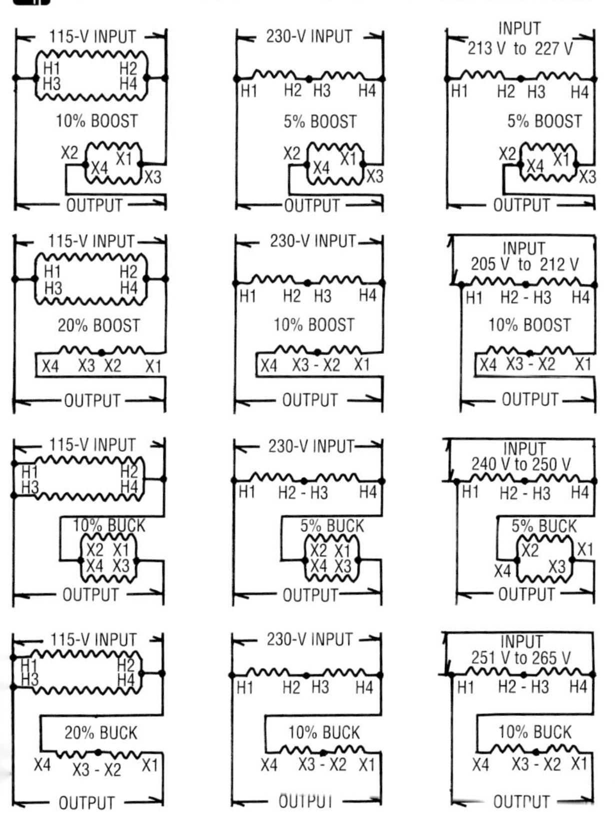

FULL-LOAD CURRENTS

THREE-PHASE TRANSFORMERS VOLTAGE (LINE TO LINE)						SINGLE-PHASE TRANSFORMERS VOLTAGE					
KVA RATING	208	240	480	2400	4160	KVA RATING	120	208	240	480	2400
3	8.3	7.2	3.6	.72	.416	1	8.33	4.81	4.17	2.08	.42
6	16.7	14.4	7.2	1.44	.83	3	25.0	14.4	12.5	6.25	1.25
9	25.0	21.7	10.8	2.17	1.25	5	41.7	24.0	20.8	10.4	2.08
15	41.6	36.1	18.0	3.6	2.08	7.5	62.5	36.1	31.3	15.6	3.13
30	83.3	72.2	36.1	7.2	4.16	10	83.3	48.1	41.7	20.8	4.17
45	124.9	108.3	54.1	10.8	6.25	15	125.0	72.1	62.5	31.3	6.25
75	208.2	180.4	90.2	18.0	10.4	25	208.3	120.2	104.2	52.1	10.4
100	277.6	240.6	120.3	24.1	13.9	37.5	312.5	180.3	156.3	78.1	15.6
150	416.4	360.9	180.4	36.1	20.8	50	416.7	240.4	208.3	104.2	20.8
225	624.6	541.3	270.6	54.1	31.2	75	625.0	360.6	312.5	156.3	31.3
300	832.7	721.7	360.9	72.2	41.6	100	833.3	480.8	416.7	208.3	41.7
500	1387.9	1202.8	601.4	120.3	69.4	125	1041.7	601.0	520.8	260.4	52.1
750	2081.9	1804.3	902.1	180.4	104.1	167.5	1395.8	805.3	697.9	349.0	69.8
1000	2775.8	2405.7	1202.8	240.6	138.8	200	1666.7	961.5	833.3	416.7	83.3
1500	4163.7	3608.5	1804.3	360.9	208.2	250	2083.3	1201.9	1041.7	520.8	104.2
2000	5551.6	4811.4	2405.7	481.1	277.6	333	2775.0	1601.0	1387.5	693.8	138.8
2500	6939.5	6014.2	3007.1	601.4	347.0	500	4166.7	2403.8	2083.3	1041.7	208.3
5000	13,879.0	12,028.5	6014.2	1202.8	694.0						
7500	20,818.5	18,042.7	9021.4	1804.3	1040.9						
10,000	27,758.0	24,057.0	12,028.5	2405.7	1387.9						

$I = \frac{KVA \times 1000}{E \times 1.73}$ OR $KVA = \frac{E \times I \times 1.73}{1000}$

$I = \frac{KVA \times 1000}{E}$ OR $KVA = \frac{E \times I}{1000}$

TRANSFORMER CALCULATIONS

To better understand the following formulas, review the rule of transposition in equations.

A multiplier may be removed from one side of an equation by making it a divisor on the other side; or a divisor may be removed from one side of an equation by making it a multiplier on the other side.

1. VOLTAGE AND CURRENT: PRIMARY (p) AND SECONDARY (s)

POWER (p) = POWER (s) OR $Ep \times Ip = Es \times Is$

A. $Ep = \dfrac{Es \times Is}{Ip}$

B. $Ip = \dfrac{Es \times Is}{Ep}$

C. $\dfrac{Ep \times Ip}{Es} = Is$

D. $\dfrac{Ep \times Ip}{Is} = Es$

2. VOLTAGE AND TURNS IN COIL:

VOLTAGE (p) x TURNS (s) = VOLTAGE (s) x TURNS (p)

OR

$Ep \times Ts = Es \times Tp$

A. $Ep = \dfrac{Es \times Tp}{Ts}$

B. $Ts = \dfrac{Es \times Tp}{Ep}$

C. $\dfrac{Ep \times Ts}{Es} = Tp$

D. $\dfrac{Ep \times Ts}{Tp} = Es$

3. AMPERES AND TURNS IN COIL:

AMPERES (p) x TURNS (p) = AMPERES (s) x TURNS (s)

OR

$Ip \times Tp = Is \times Ts$

A. $Ip = \dfrac{Is \times Ts}{Tp}$

B. $Tp = \dfrac{Is \times Ts}{Ip}$

C. $\dfrac{Ip \times Tp}{Is} = Ts$

D. $\dfrac{Ip \times Tp}{Ts} = Is$

SIZING TRANSFORMERS

Single-Phase Transformers

Size a 480-volt, primary, or 240/120-volt, secondary, single-phase transformer for the following single-phase incandescent lighting load consisting of 48 recessed fixtures each rated 2 amperes, 120 volts. Each fixture has a 150-watt lamp.
*(These fixtures can be evenly balanced on the transformer.)

Find total volt-amperes using fixture ratings -
do not use lamp watt rating.

2 amperes x 120 volts = 240 volt-amperes
240 VA x 48 = 11,520 VA
11,520 VA/1000 = 11.52 KVA

The single-phase transformer that meets or exceeds this value is **15 KVA**.

*24 lighting fixtures would be connected line one to the common neutral, and 24 lighting fixtures would be connected line two to the common neutral.

Three-Phase Transformers

Size a 480-volt, primary, or 240-volt, secondary, 3-phase transformer (polyphase unit) to supply one 3-phase, 25-KVA process heater and one single-phase, 5-KW unit heater.

The 5-KW unit heater cannot be balanced across all three phases. The 5 KW will be on one phase only. Adding the loads directly will undersize the transformer. Common practice is to put an imaginary load equal to the single-phase load on the other two phases.
5 KW x 3 = 15 KW*
The 25 KVA is three phase; use 25 KVA.
25 KVA + 15 KVA* = 40 KVA
The nearest 3-phase transformer that meets or exceeds this value is a **45 KVA**.
*(KVA = KW at unity power factor) (Transformers are rated in KVA.)

SINGLE-PHASE TRANSFORMER

Primary and Secondary Amperes

A 480/240-volt, single-phase, 50-KVA transformer (Z = 2%) is to be installed. Calculate primary and secondary amperes and short-circuit amperes.

Primary Amperes:

$$I_p = \frac{KVA \times 1000}{E_p} = \frac{50 \times 1000}{480} = \frac{50{,}000}{480} = \textbf{104 AMPERES}$$

Secondary Amperes:

$$I_s = \frac{KVA \times 1000}{E_s} = \frac{50 \times 1000}{240} = \frac{50{,}000}{240} = \textbf{208 AMPERES}$$

Short-Circuit Amperes:*

$$I_{sc} = \frac{I_s}{\%Z} = \frac{208}{0.02} = \textbf{10,400 AMPERES}$$

THREE-PHASE TRANSFORMER

Primary and Secondary Amperes

A 480/208-volt, 3-phase, 100-KVA transformer (Z = 1%) is to be installed. Calculate primary and secondary amperes and short-circuit amperes.

Primary Amperes:

$$I_p = \frac{KVA \times 1000}{E_p \times 1.73} = \frac{100 \times 1000}{480 \times 1.73} = \frac{100{,}000}{831} = \textbf{120 AMPERES}$$

Secondary Amperes:

$$I_s = \frac{KVA \times 1000}{E_s \times 1.73} = \frac{100 \times 1000}{208 \times 1.73} = \frac{100{,}000}{360} = \textbf{278 AMPERES}$$

Short-Circuit Amperes:*

$$I_{sc} = \frac{I_s}{\%Z} = \frac{278}{0.01} = \textbf{27,800 AMPERES}$$

*Short-circuit amperes is the current that would flow if the transformers' secondary terminals were shorted phase to phase. See *Ugly's* pages 53–56 for calculating short-circuit amperes point to point using the Cooper Bussmann method.

THREE-PHASE CONNECTIONS

Wye (Star)

Voltage from "A", "B", or "C" to Neutral = E_{PHASE} (E_P)

Voltage between A-B, B-C, or C-A = E_{LINE} (E_L)

$I_L = I_P$, if balanced.

If unbalanced,

$I_N = \sqrt{I_A^2 + I_B^2 + I_C^2 - (I_A \times I_B) - (I_B \times I_C) - (I_C \times I_A)}$

$E_L = Ep \times 1.73$

$E_P = E_L \div 1.73$

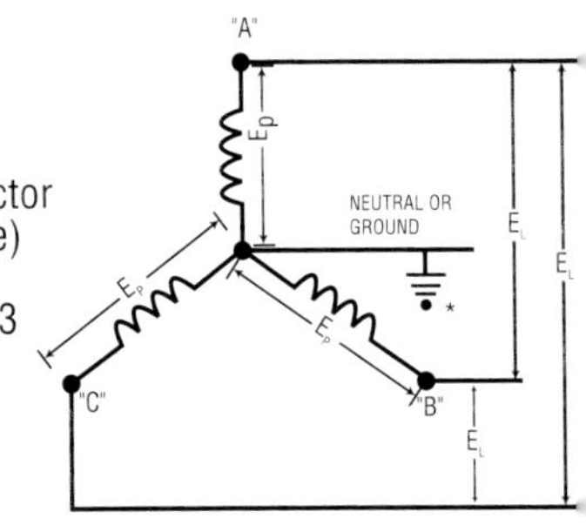

(True Power)
Power =

$I_L \times E_L \times 1.73 \times$ Power Factor (cosine)

(Apparent Power)
Volt-Amperes = $I_L \times E_L \times 1.73$

Delta

E_{LINE} (E_L) = E_{PHASE} (E_P)

$I_{LINE} = I_P \times 1.73$

$I_{PHASE} = I_L \div 1.73$

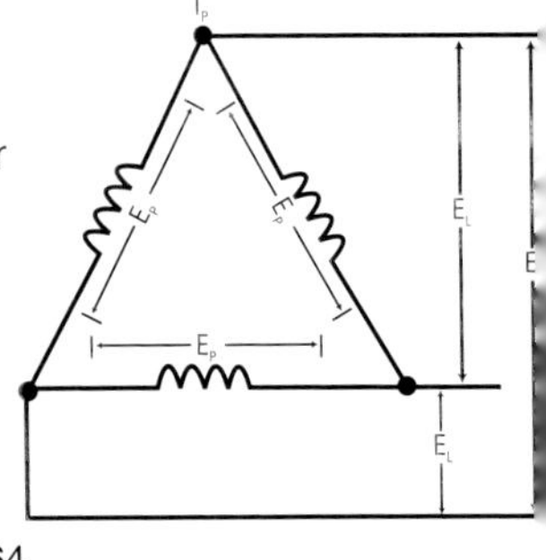

(True Power)
Power =

$I_L \times E_L \times 1.73 \times$ Power Factor (cosine)

(Apparent Power)
Volt-Amperes = $I_L \times E_L \times 1.73$

*Neutral could be ungrounded. Also see *NEC* Article 250, Grounding and Bonding.

THREE-PHASE STANDARD PHASE ROTATION

Transformers

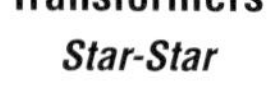

Star-Delta

ADDITIVE POLARITY
30° ANGULAR DISPLACEMENT

Star-Star

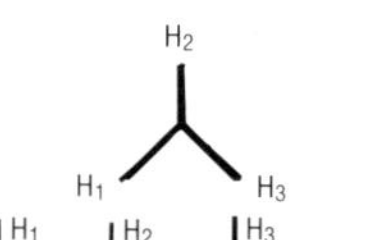

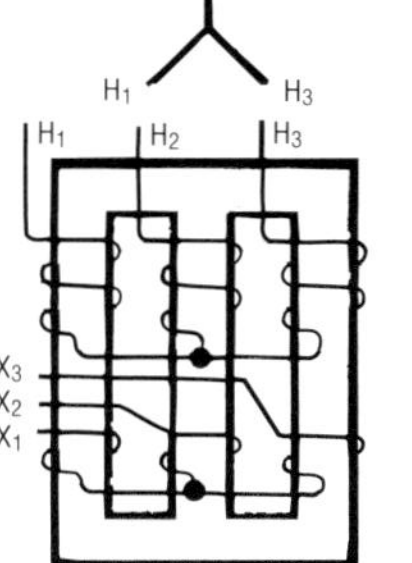

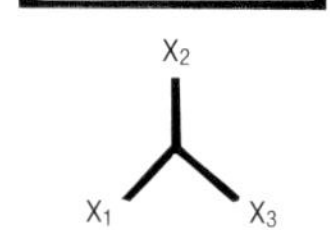

SUBTRACTIVE POLARITY
0° PHASE DISPLACEMENT

Delta-Delta

SUBTRACTIVE POLARITY
0° PHASE DISPLACEMENT

TRANSFORMER CONNECTIONS

Series Connections of Low-Voltage Windings

Delta-Delta

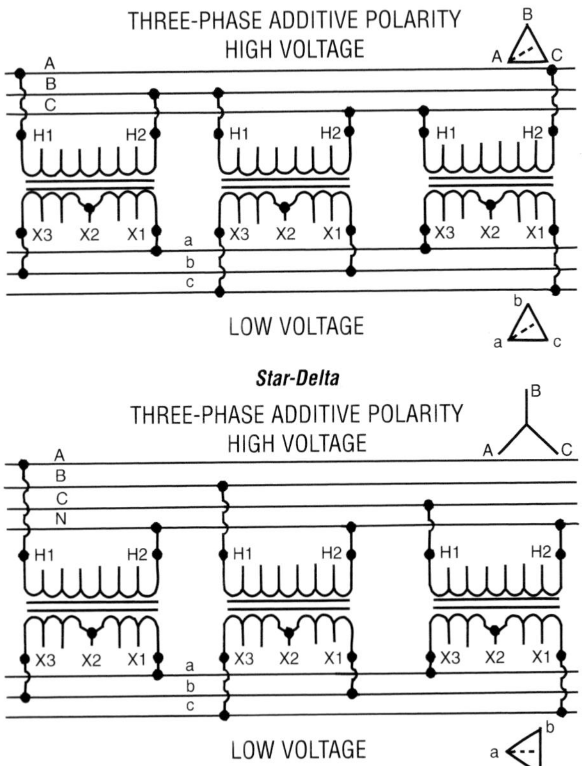

NOTE: Single-phase transformers should be thoroughly checked for impedance, polarity, and voltage ratio before installation.

TRANSFORMER CONNECTIONS

Series Connections of Low-Voltage Windings

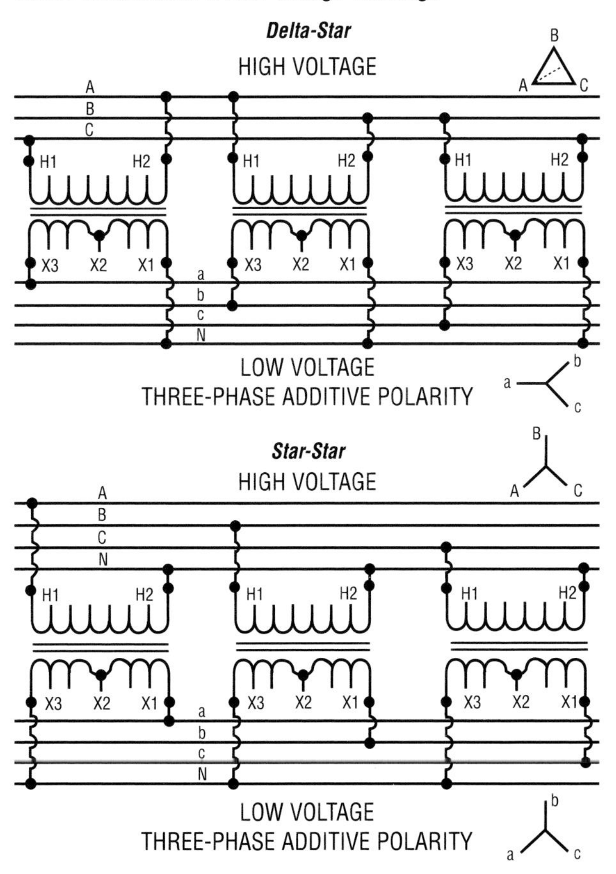

NOTE: For additive polarity the H1 and the X1 bushings are diagonally opposite each other.

TRANSFORMER CONNECTIONS

Series Connections of Low-Voltage Windings

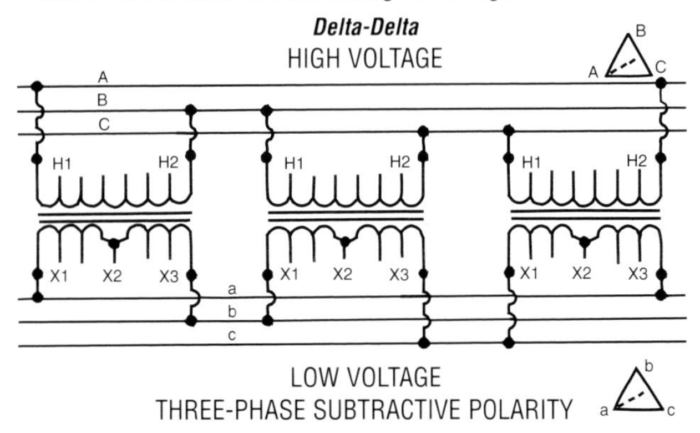

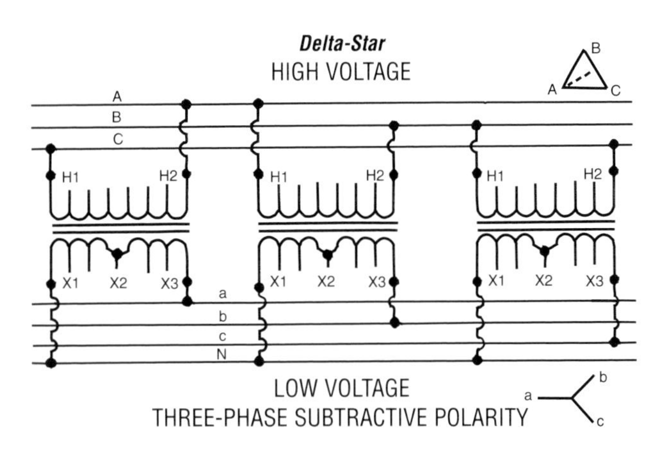

NOTE: For subtractive polarity the H1 and the X1 bushings are directly opposite each other.

MISCELLANEOUS WIRING DIAGRAMS

Two Three-Way Switches

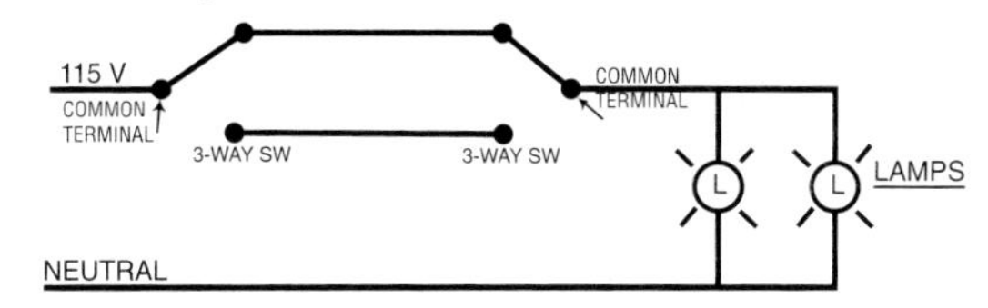

Two Three-Way Switches and One Four-Way Switch

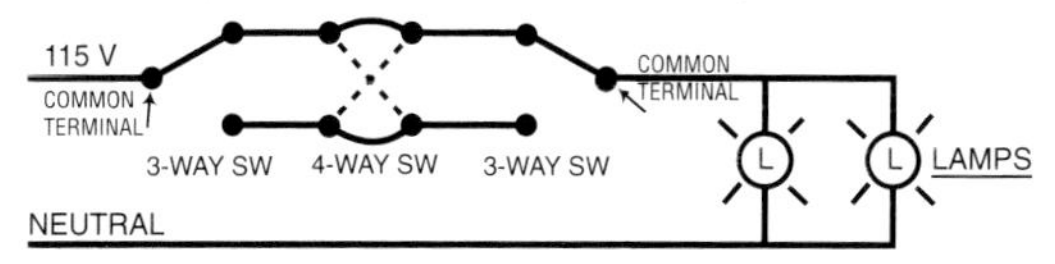

Bell Circuit

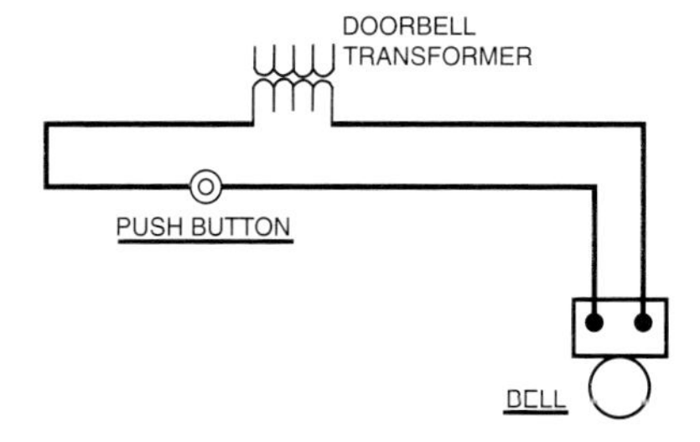

MISCELLANEOUS WIRING DIAGRAMS

Remote-Control Circuit: One Relay and One Switch

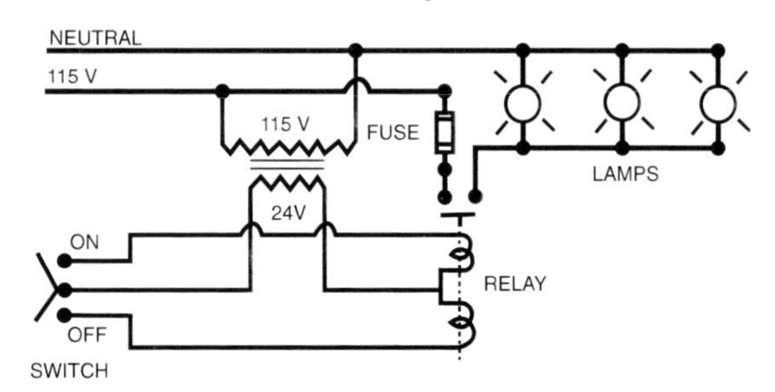

SUPPORTS FOR RIGID METAL CONDUIT

CONDUIT SIZE	DISTANCE BETWEEN SUPPORTS
½" - ¾"	10 FEET
1"	12 FEET
1¼" - 1½"	14 FEET
2" - 2½"	16 FEET
3" AND LARGER	20 FEET

Source: NFPA 70, *National Electrical Code®*, NFPA, Quincy, MA, 2013, Table 344.30(B)(2), as modified.

SUPPORTS FOR RIGID NONMETALLIC CONDUIT

CONDUIT SIZE	DISTANCE BETWEEN SUPPORTS
½" - 1"	3 FEET
1¼" - 2"	5 FEET
2½" - 3"	6 FEET
3½" - 5"	7 FEET
6	8 FEET

For SI units: (Supports) 1 foot = 0.3048 meter

Source: NFPA 70, *National Electrical Code®*, NFPA, Quincy, MA, 2013, Table 352.30, as modified.

CONDUCTOR PROPERTIES

Size	Area	Conductors				DC Resistance at 75°C (167°F)		
		Stranding		Overall		Copper		Aluminum
AWG/ kcmil	Cir. Mils	Quantity	Diam. (in.)	Diam. (in.)	Area (in.^{2})	Uncoated ohm/kFT	Coated ohm/kFT	ohm/ kFT
18	1620	1	----	0.040	0.001	7.77	8.08	12.8
18	1620	7	0.015	0.046	0.002	7.95	8.45	13.1
16	2580	1	----	0.051	0.002	4.89	5.08	8.05
16	2580	7	0.019	0.058	0.003	4.99	5.29	8.21
14	4110	1	----	0.064	0.003	3.07	3.19	5.06
14	4110	7	0.024	0.073	0.004	3.14	3.26	5.17
12	6530	1	----	0.081	0.005	1.93	2.01	3.18
12	6530	7	0.030	0.092	0.006	1.98	2.05	3.25
10	10380	1	----	0.102	0.008	1.21	1.26	2.00
10	10380	7	0.038	0.116	0.011	1.24	1.29	2.04
8	16510	1	----	0.128	0.013	0.764	0.786	1.26
8	16510	7	0.049	0.146	0.017	0.778	0.809	1.28
6	26240	7	0.061	0.184	0.027	0.491	0.510	0.808
4	41740	7	0.077	0.232	0.042	0.308	0.321	0.508
3	52620	7	0.087	0.260	0.053	0.245	0.254	0.403
2	66360	7	0.097	0.292	0.067	0.194	0.201	0.319
1	83690	19	0.066	0.332	0.087	0.154	0.160	0.253
1/0	105600	19	0.074	0.372	0.109	0.122	0.127	0.201
2/0	133100	19	0.084	0.418	0.137	0.0967	0.101	0.159
3/0	167800	19	0.094	0.470	0.173	0.0766	0.0797	0.126
4/0	211600	19	0.106	0.528	0.219	0.0608	0.0626	0.100
250	----	37	0.082	0.575	0.260	0.0515	0.0535	0.0847
300	----	37	0.090	0.630	0.312	0.0429	0.0446	0.0707
350	----	37	0.097	0.681	0.364	0.0367	0.0382	0.0605
400	----	37	0.104	0.728	0.416	0.0321	0.0331	0.0529
500	----	37	0.116	0.813	0.519	0.0258	0.0265	0.0424
600	----	61	0.099	0.893	0.626	0.0214	0.0223	0.0353
700	----	61	0.107	0.964	0.730	0.0184	0.0189	0.0303
750	----	61	0.111	0.998	0.782	0.0171	0.0176	0.0282
800	----	61	0.114	1.030	0.834	0.0161	0.0166	0.0265
900	----	61	0.122	1.094	0.940	0.0143	0.0147	0.0235
1000	----	61	0.128	1.152	1.042	0.0129	0.0132	0.0212
1250	----	91	0.117	1.289	1.305	0.0103	0.0106	0.0169
1500	----	91	0.128	1.412	1.566	0.00858	0.00883	0.0141
1750	----	127	0.117	1.526	1.829	0.00735	0.00756	0.0121
2000	----	127	0.126	1.632	2.092	0.00643	0.00662	0.0106

These resistance values are valid ONLY for the parameters as given. Using conductors having coated strands, different stranding type, and, especially, other temperatures changes the resistance.

Equation for temperature change: $R_2 = R_1\ [1 + \alpha(T_2 - 75)]$ where $\alpha_{cu} = 0.00323$, $\alpha_{AL} = 0.00330$ at 75°C (167°F).

See *NEC* Chapter 9, Table 8. See *Ugly's* pages 133–141 for metric conversions.

Source: NFPA 70, *National Electrical Code*®, NFPA, Quincy, MA, 2013, Table 8, as modified.

AC RESISTANCE AND REACTANCE FOR 600-VOLT CABLES, THREE-PHASE, 60-HZ, 75°C (167°F): THREE SINGLE CONDUCTORS IN CONDUIT

	Ohms to Neutral per 1000 Feet														
Size	X_L (Reactance) for All Wires		AC Resistance for Uncoated Copper Wires			AC Resistance for Aluminum Wires			Effective Z at 0.85 PF for Uncoated Copper Wires			Effective Z at 0.85 PF for Aluminum Wires			Size
AWG/ kcmil	PVC, Al. Conduits	Steel Conduit	PVC Conduit	Al. Conduit	Steel Conduit	PVC Conduit	Al. Conduit	Steel Conduit	PVC Conduit	Al. Conduit	Steel Conduit	PVC Conduit	Al. Conduit	Steel Conduit	AWG/ kcmil
14	0.058	0.073	3.1	3.1	3.1	–	–	–	2.7	2.7	2.7	–	–	–	14
12	0.054	0.068	2.0	2.0	2.0	3.2	3.2	3.2	1.7	1.7	1.7	2.8	2.8	2.8	12
10	0.050	0.063	1.2	1.2	1.2	2.0	2.0	2.0	1.1	1.1	1.1	1.8	1.8	1.8	10
8	0.052	0.065	0.78	0.78	0.78	1.3	1.3	1.3	0.69	0.69	0.70	1.1	1.1	1.1	8
6	0.051	0.064	0.49	0.49	0.49	0.81	0.81	0.81	0.44	0.45	0.45	0.71	0.72	0.72	6
4	0.048	0.060	0.31	0.31	0.31	0.51	0.51	0.51	0.29	0.29	0.30	0.46	0.46	0.46	4
3	0.047	0.059	0.25	0.25	0.25	0.40	0.41	0.40	0.23	0.24	0.24	0.37	0.37	0.37	3
2	0.045	0.057	0.19	0.20	0.20	0.32	0.32	0.32	0.19	0.19	0.20	0.30	0.30	0.30	2
1	0.046	0.057	0.15	0.16	0.16	0.25	0.26	0.25	0.16	0.16	0.16	0.24	0.24	0.25	1
1/0	0.044	0.055	0.12	0.13	0.12	0.20	0.21	0.20	0.13	0.13	0.13	0.19	0.20	0.20	1/0
2/0	0.043	0.054	0.10	0.10	0.10	0.16	0.16	0.16	0.11	0.11	0.11	0.16	0.16	0.16	2/0
3/0	0.042	0.052	0.077	0.082	0.079	0.13	0.13	0.13	0.088	0.092	0.094	0.13	0.13	0.14	3/0
4/0	0.041	0.051	0.062	0.067	0.063	0.10	0.11	0.10	0.074	0.078	0.080	0.11	0.11	0.11	4/0
250	0.041	0.052	0.052	0.057	0.054	0.085	0.090	0.086	0.066	0.070	0.073	0.094	0.098	0.10	250
300	0.041	0.051	0.044	0.049	0.045	0.071	0.076	0.072	0.059	0.063	0.065	0.082	0.086	0.088	300
350	0.040	0.050	0.038	0.043	0.039	0.061	0.066	0.063	0.053	0.058	0.060	0.073	0.077	0.080	350
400	0.040	0.049	0.033	0.038	0.035	0.054	0.059	0.055	0.049	0.053	0.056	0.066	0.071	0.073	400
500	0.039	0.048	0.027	0.032	0.029	0.043	0.048	0.045	0.043	0.048	0.050	0.057	0.061	0.064	500
600	0.039	0.048	0.023	0.028	0.025	0.036	0.041	0.038	0.040	0.044	0.047	0.051	0.055	0.058	600
750	0.038	0.048	0.019	0.024	0.021	0.029	0.034	0.031	0.036	0.040	0.043	0.045	0.049	0.052	750
1000	0.037	0.046	0.015	0.019	0.018	0.023	0.027	0.025	0.032	0.036	0.040	0.039	0.042	0.046	1000

NOTES: See *NEC* Table 9, pages 722–723 for assumptions and explanations. See *Ugly's* pages 133–141 for metric conversions.
Source: NFPA 70, *National Electrical Code®*, NFPA, Quincy, MA, 2013, Table 9, as modified.

ALLOWABLE AMPACITIES OF CONDUCTORS: RACEWAY, CABLE, OR EARTH

Allowable Ampacities of Insulated Conductors Rated Up to and Including 2000 Volts, 60°C Through 90°C (140°F Through 194°F), **Not More Than Three Current-Carrying Conductors** in **Raceway, Cable, or Earth** (Directly Buried), Based on Ambient Air Temperature of **30°C** (86°F)*

SIZE	60°C (140°F)	75°C (167°F)	90°C (194°F)	60°C (140°F)	75°C (167°F)	90°C (194°F)	SIZE
AWG or kcmil	**TYPES** TW, UF	**TYPES** RHW, THHW, THW, THWN, XHHW, USE, ZW	**TYPES** TBS, SA, SIS, FEP, FEPB, MI, RHH, RHW-2, THHN, THHW, THW-2,THWN-2, USE-2, XHH, XHHW, XHHW-2, ZW-2	**TYPES** TW, UF	**TYPES** RHW, THHW, THW, THWN, XHHW, USE	**TYPES** TBS, SA, SIS, THHN, THHW, THW-2, THWN-2, RHH, RHW-2, USE-2, XHH, XHHW, XHHW-2, ZW-2	AWG or kcmil
		COPPER		**ALUMINUM OR COPPER-CLAD ALUMINUM**			
14**	15	20	25	----	----	----	----
12**	20	25	30	15	20	25	12**
10**	30	35	40	25	30	35	10**
8	40	50	55	35	40	45	8
6	55	65	75	40	50	55	6
4	70	85	95	55	65	75	4
3	85	100	115	65	75	85	3
2	95	115	130	75	90	100	2
1	110	130	145	85	100	115	1
1/0	125	150	170	100	120	135	1/0
2/0	145	175	195	115	135	150	2/0
3/0	165	200	225	130	155	175	3/0
4/0	195	230	260	150	180	205	4/0
250	215	255	290	170	205	230	250
300	240	285	320	195	230	260	300
350	260	310	350	210	250	280	350
400	280	335	380	225	270	305	400
500	320	380	430	260	310	350	500
600	350	420	475	285	340	385	600
700	385	460	520	315	375	425	700
750	400	475	535	320	385	435	750
800	410	490	555	330	395	445	800
900	435	520	585	355	425	480	900
1000	455	545	615	375	445	500	1000
1250	495	590	665	405	485	545	1250
1500	525	625	705	435	520	585	1500
1750	545	650	735	455	545	615	1750
2000	555	665	750	470	560	630	2000

*Refer to 310.15(B)(2) for the ampacity correction factors where the ambient temperature is other than 30°C (86°F).

**Refer to 240.4(D) for conductor overcurrent protection limitations.

See *NEC* Ambient Temperatures Correction Table 310.15(B)(2)(a).

See *Ugly's* page 77 for Adjustment Examples.

Source: NFPA 70, *National Electrical Code*®, NFPA, Quincy, MA, 2013, Table 310.15(B)(16), as modified.

ALLOWABLE AMPACITIES OF CONDUCTORS: FREE AIR

Allowable Ampacities of **Single-Insulated Conductors** Rated Up to and Including 2000 Volts in Free Air, Based on Ambient Air Temperature of **30°C** (86°F)*

SIZE	60°C (140°F)	75°C (167°F)	90°C (194°F)	60°C (140°F)	75°C (167°F)	90°C (194°F)	SIZE
AWG or kcmil	**TYPES** TW, UF	**TYPES** RHW, THHW, THW, THWN, XHHW, USE, ZW	**TYPES** TBS, SA, SIS, FEP, FEPB, MI, RHH, RHW-2, THHN, THHW, THW-2, THWN-2, USE-2, XHH, XHHW, XHHW-2, ZW-2	**TYPES** TW, UF	**TYPES** RHW, THHW, THW, THWN, XHHW, USE	**TYPES** TBS, SA, SIS, THHN, THHW, THW-2, THWN-2, RHH, RHW-2, USE-2, XHH, XHHW, XHHW-2, ZW-2	AWG or kcmil
	COPPER			**ALUMINUM OR COPPER-CLAD ALUMINUM**			
14**	25	30	35	----	----	----	----
12**	30	35	40	25	30	35	12**
10**	40	50	55	35	40	45	10**
8	60	70	80	45	55	60	8
6	80	95	105	60	75	85	6
4	105	125	140	80	100	115	4
3	120	145	165	95	115	130	3
2	140	170	190	110	135	150	2
1	165	195	220	130	155	175	1
1/0	195	230	260	150	180	205	1/0
2/0	225	265	300	175	210	235	2/0
3/0	260	310	350	200	240	270	3/0
4/0	300	360	405	235	280	315	4/0
250	340	405	455	265	315	355	250
300	375	445	500	290	350	395	300
350	420	505	570	330	395	445	350
400	455	545	615	355	425	480	400
500	515	620	700	405	485	545	500
600	575	690	780	455	540	615	600
700	630	755	850	500	595	670	700
750	655	785	885	515	620	700	750
800	680	815	920	535	645	725	800
900	730	870	980	580	700	790	900
1000	780	935	1055	625	750	845	1000
1250	890	1065	1200	710	855	965	1250
1500	980	1175	1325	795	950	1070	1500
1750	1070	1280	1445	875	1050	1185	1750
2000	1155	1385	1560	960	1150	1295	2000

*Refer to 310.15(B)(2) for the ampacity correction factors where the ambient temperature is other than 30°C (86°F).
**Refer to 240.4(D) for conductor overcurrent protection limitations.
See *NEC* Ambient Temperatures Correction Table 310.15(B)(2)(a).
See *Ugly's* page 77 for Adjustment Examples.
Source: NFPA 70, *National Electrical Code®*, NFPA, Quincy, MA, 2013, Table 310.15(B)(17), as modified.

ALLOWABLE AMPACITIES OF CONDUCTORS: RACEWAY OR CABLE

Allowable Ampacities of Insulated Conductors Rated Up to and Including 2000 Volts, 150°C Through 250°C (302°F Through 482°F), **Not More Than Three Current-Carrying Conductors** in **Raceway or Cable,** Based on Ambient Air Temperature of **40°C** (104°F)*

SIZE	150°C (302°F)	200°C (392°F)	250°C (482°F)	150°C (302°F)	SIZE
AWG or kcmil	**TYPE** Z	**TYPES** FEP, FEPB, PFA, SA	**TYPES** PFAH, TFE	**TYPE** Z	AWG or kcmil
	COPPER		**NICKEL or NICKEL-COATED COPPER**	**ALUMINUM or COPPER-CLAD ALUMINUM**	
14	34	36	39	---	14
12	43	45	54	30	12
10	55	60	73	44	10
8	76	83	93	57	8
6	96	110	117	75	6
4	120	125	148	94	4
3	143	152	166	109	3
2	160	171	191	124	2
1	186	197	215	145	1
1/0	215	229	244	169	1/0
2/0	251	260	273	198	2/0
3/0	288	297	308	227	3/0
4/0	332	346	361	260	4/0

*Refer to 310.15(B)(2) for the ampacity correction factors where the ambient temperature is other than 40°C (104°F). See *NEC* Ambient Temperatures Correction Table 310.15(B)(2)(b).
Source: NFPA 70, *National Electrical Code®*, NFPA, Quincy, MA, 2013, Table 310.15(B)(18).

ALLOWABLE AMPACITIES OF CONDUCTORS: FREE AIR

Allowable Ampacities of **Single-Insulated Conductors**, Rated Up to and Including 2000 Volts, 150°C Through 250°C (302°F Through 482°F), **in Free Air**, Based on Ambient Air Temperature of **40°C** (104°F)*

SIZE	150°C (302°F)	200°C (392°F)	250°C (482°F)	150°C (302°F)	SIZE
AWG or kcmil	**TYPE** Z	**TYPES** FEP, FEPB, PFA, SA	**TYPES** PFAH, TFE	**TYPE** Z	AWG or kcmil
	COPPER		**NICKEL or NICKEL-COATED COPPER**	**ALUMINUM or COPPER-CLAD ALUMINUM**	
14	46	54	59	---	14
12	60	68	78	47	12
10	80	90	107	63	10
8	106	124	142	83	8
6	155	165	205	112	6
4	190	220	278	148	4
3	214	252	327	170	3
2	255	293	381	198	2
1	293	344	440	228	1
1/0	339	399	532	263	1/0
2/0	390	467	591	305	2/0
3/0	451	546	708	351	3/0
4/0	529	629	830	411	4/0

*Refer to 310.15(B)(2) for the ampacity correction factors where the ambient temperature is other than 40°C (104°F). See *NEC* Ambient Temperatures Correction Table 310.15(B)(2)(b).

Source: NFPA 70, *National Electrical Code®*, NFPA, Quincy, MA, 2013, Table 310.15(B)(19).

AMPACITY CORRECTION AND ADJUSTMENT FACTORS

Examples

Ugly's page 73 shows ampacity values for not more than three current-carrying conductors in a raceway or cable and an ambient (surrounding) temperature of 30°C (86°F).

Example 1: A raceway contains three #3 THWN conductors for a three-phase circuit at an ambient temperature of 30°C (86°F).
Ugly's page 73, 75°C (167°F) column indicates **100 amperes**.

Example 2: A raceway contains three #3 THWN conductors for a three-phase circuit at an ambient temperature of 40°C (104°F). *Ugly's* page 73, 75°C (167°F) column indicates **100 amperes**. This value must be corrected for ambient temperature (see note on Temperature Correction Factors at bottom of *Ugly's* page 73). 40°C (104°F) factor is **0.88**.
100 amperes x 0.88 = **88 amperes** = corrected ampacity

Example 3: A raceway contains six #3 THWN conductors for two three-phase circuits at an ambient temperature of 30°C (86°F).
Ugly's page 73, 75°C (167°F) column indicates **100 amperes**. This value must be adjusted for more than three current-carrying conductors. The table on *Ugly's* page 78 requires an adjustment of **80%** for four through six current-carrying conductors.
100 amperes x 80% = **80 amperes**
The adjusted ampacity is **80 amperes**

Example 4: A raceway contains six #3 THWN conductors for two three-phase circuits in an ambient temperature of 40°C (104°F). These conductors must be derated for both temperature and number of conductors.
Ugly's page 73, 75°C (167°F) column indicates **100 amperes**
NEC Table 310.15(B)(2)(a), 40°C (104°F) temperature factor is **0.88**
Ugly's page 78, 4 - 6 conductor factor is **0.80**
100 amperes x 0.88 x 0.80 = **70.4 amperes**
The new derated ampacity is **70.4 amperes**.

ADJUSTMENT FACTORS

For More Than Three Current-Carrying Conductors in a Raceway or Cable

Number of Current-Carrying Conductors*	Percent of Values in Tables 310.15(B)(16) through 310.15(B)(19) as Adjusted for Ambient Temperature if Necessary
4–6	80
7–9	70
10–20	50
21–30	45
31–40	40
41 and above	35

*Number of conductors is the total number of conductors, spares included, in the raceway or cable adjusted in accordance with 310.15(B)(5) and (6).

Source: NFPA 70, *National Electrical Code®*, NFPA, Quincy, MA, 2013, Table 310.15(B)(3)(a).

Conductor and Equipment Termination Ratings*

Examples

A 150-ampere circuit breaker is labeled for 75°C (167°F) terminations and is selected to be used for a 150-ampere, noncontinuous load. It would be permissible to use a 1/0 THWN conductor that has a 75°C (167°F) insulation rating and has an ampacity of 150 amperes (*Ugly's* page 73).

When a THHN (90°C [194°F]) conductor is connected to a 75°C (167°F) termination, it is limited to the 75°C (167°F) ampacity. Therefore, if a 1 THHN conductor with a rating of 150 amperes were connected to a 75°C (167°F) terminal, its ampacity would be limited to 130 amperes instead of 150 amperes, which is too small for the load (*Ugly's* page 73).

If the 150-ampere, noncontinuous load listed above uses 1/0 THWN conductors rated 150 amperes and the conductors are in an ambient temperature of 40°C (104°F), the conductors would have to be corrected for the ambient temperature.

ADJUSTMENT FACTORS

From *NEC* Table 310.15(B)(2)(a), 30°C (86°F) ambient temperature correction factor = **0.88**
1/0 THWN = 150 amperes
150 amperes x 0.88 = **132 amperes** (which is too small for the load, so a larger size conductor is required).

Apply temperature correction factors to the next size THWN conductor.
2/0 THWN = 175 amperes (from the 75°C [167°F] column - *Ugly's* page 73)
175 amperes x 0.88 = **154 amperes**. This size is suitable for the 150-ampere load.

The advantage of using 90°C (194°F) conductors is that you can apply ampacity derating factors to the higher 90°C (194°F) ampacity rating, and it may save you from going to a larger conductor.

1/0 THHN = 170 amperes (from the 90°C [194°F] column - *Ugly's* page 73).
30°C [86°F] ambient temperature correction factor = .91 (*NEC* Table 310.15(B)(2)(a))
170 amperes x 0.91 = **154.7 amperes**.
This size is suitable for the 150-ampere load.
This 90°C [194°F] conductor can be used but can never have a final derated ampacity over the rating of 1/0 THWN 75°C [167°F] rating of 150 amperes.

You are allowed to use higher temperature (insulated conductors) such as THHN (90°C [194°F]) conductors on 60°C (140°F) or 75°C (167°F) terminals of circuit breakers and equipment, and you are allowed to derate from the higher value for temperature and number of conductors, but the final derated ampacity is limited to the 60°C or 75°C (140°F or 167°F) terminal insulation labels.

*See *NEC* 2014 Article 110.14(C)(1) and (2).

CONDUCTOR APPLICATIONS AND INSULATIONS

TRADE NAME	LETTER	MAX. TEMP.	APPLICATION PROVISIONS
FLUORINATED ETHYLENE PROPYLENE	FEP OR FEPB	90°C (194°F)	DRY AND DAMP LOCATIONS
		200°C (392°F)	DRY LOCATIONS - SPECIAL APPLICATIONS[1]
MINERAL INSULATION (METAL SHEATHED)	MI	90°C (194°F)	DRY AND WET LOCATIONS
		250°C (482°F)	SPECIAL APPLICATIONS[1]
MOISTURE-, HEAT-, AND OIL-RESISTANT THERMOPLASTIC	MTW	60°C (140°F)	MACHINE TOOL WIRING IN WET LOCATIONS
		90°C (194°F)	MACHINE TOOL WIRING IN DRY LOCATIONS, Informational Note: See NFPA 79
PAPER		85°C (185°F)	FOR UNDERGROUND SERVICE CONDUCTORS, OR BY SPECIAL PERMISSION
PERFLUORO-ALKOXY	PFA	90°C (194°F)	DRY AND DAMP LOCATIONS
		200°C (392°F)	DRY LOCATIONS - SPECIAL APPLICATIONS[1]
PERFLUORO-ALKOXY	PFAH	250°C (482°F)	DRY LOCATIONS ONLY. ONLY FOR LEADS WITHIN APPARATUS OR WITHIN RACEWAYS CONNECTED TO APPARATUS (NICKEL OR NICKEL-COATED COPPER ONLY)
THERMOSET	RHH	90°C (194°F)	DRY AND DAMP LOCATIONS
MOISTURE-RESISTANT THERMOSET	RHW	75°C (167°F)	DRY AND WET LOCATIONS
MOISTURE-RESISTANT THERMOSET	RHW-2	90°C (194°F)	DRY AND WET LOCATIONS
SILICONE	SA	90°C (194°F)	DRY AND DAMP LOCATIONS
		200°C (392°F)	SPECIAL APPLICATIONS[1]
THERMOSET	SIS	90°C (194°F)	SWITCHBOARD AND SWITCHGEAR WIRING ONLY
THERMOPLASTIC AND FIBROUS OUTER BRAID	TBS	90°C (194°F)	SWITCHBOARD AND SWITCHGEAR WIRING ONLY

See *Ugly's* page 82 for footnotes on special provisions and/or applications.

CONDUCTOR APPLICATIONS AND INSULATIONS

TRADE NAME	LETTER	MAX. TEMP.	APPLICATION PROVISIONS
EXTENDED POLYTETRAFLUORO-ETHYLENE	TFE	250°C (482°F)	DRY LOCATIONS ONLY, ONLY FOR LEADS WITHIN APPARATUS OR WITHIN RACEWAYS CONNECTED TO APPARATUS, OR AS OPEN WIRING (NICKEL OR NICKEL-COATED COPPER ONLY)
HEAT-RESISTANT THERMOPLASTIC	THHN	90°C (194°F)	DRY AND DAMP LOCATIONS
MOISTURE- AND HEAT-RESISTANT THERMOPLASTIC	THHW	75°C (167°F)	WET LOCATION
		90°C (194°F)	DRY LOCATION
MOISTURE- AND HEAT-RESISTANT THERMOPLASTIC	THW	75°C (167°F)	DRY AND WET LOCATIONS
		90°C (194°F)	SPECIAL APPL. WITHIN ELECTRIC DISCHARGE LIGHTING EQUIPMENT, LIMITED TO 1000 OPEN-CIRCUIT VOLTS OR LESS, (SIZE 14-8 ONLY AS PERMITTED IN SECTION 410.68)
	THW-2	90°C (194°F)	DRY AND WET LOCATIONS
MOISTURE- AND HEAT-RESISTANT THERMOPLASTIC	THWN	75°C (167°F)	DRY AND WET LOCATIONS
	THWN-2	90°C (194°F)	
MOISTURE-RESISTANT THERMOPLASTIC	TW	60°C (140°F)	DRY AND WET LOCATIONS
UNDERGROUND FEEDER AND BRANCH-CIRCUIT CABLE SINGLE CONDUCTOR (FOR TYPE "UF" CABLE EMPLOYING MORE THAN 1 CONDUCTOR, (SEE *NEC* ARTICLE 340)	UF	60°C (140°F) 75°C (167°F)[2]	SEE *NEC* ARTICLE 340

See *Ugly's* page 82 for footnotes on special provisions and/or applications.

(continues)

CONDUCTOR APPLICATIONS AND INSULATIONS

TRADE NAME	LETTER	MAX. TEMP.	APPLICATION PROVISIONS
UNDERGROUND SERVICE-ENTRANCE CABLE SINGLE CONDUCTOR (FOR TYPE "USE" CABLE EMPLOYING MORE THAN 1 CONDUCTOR, SEE *NEC* ARTICLE 338)	USE	75°C (167°F)[2]	SEE *NEC* ARTICLE 338
	USE-2	90°C (194°F)	DRY AND WET LOCATIONS
THERMOSET	XHH	90°C (194°F)	DRY AND DAMP LOCATIONS
MOISTURE-RESISTANT THERMOSET	XHHW	90°C (194°F)	DRY AND DAMP LOCATIONS
		75°C (167°F)	WET LOCATIONS
MOISTURE-RESISTANT THERMOSET	XHHW-2	90°C (194°F)	DRY AND WET LOCATIONS
MODIFIED ETHYLENE TETRAFLUORO-ETHYLENE	Z	90°C (194°F)	DRY AND DAMP LOCATIONS
		150°C (302°F)	DRY LOCATIONS - SPECIAL APPLICATIONS[1]
MODIFIED ETHYLENE TETRAFLUORO-ETHYLENE	ZW	75°C (167°F)	WET LOCATIONS
		90°C (194°F)	DRY AND DAMP LOCATIONS
		150°C (302°F)	DRY LOCATIONS - SPECIAL APPLICATIONS[1]
	ZW-2	90°C (194°F)	DRY AND WET LOCATIONS

FOOTNOTES:

1 For signaling circuits permitting 300-volt insulation.

2 For ampacity limitation, see Section 340.80 *NEC*.

Source: NFPA 70, *National Electrical Code*®, NFPA, Quincy, MA, 2013, Table 310.104(A), as modified.

NOTE: Some insulations do not require an outer covering.

MAXIMUM NUMBER OF CONDUCTORS IN TRADE SIZES OF CONDUIT OR TUBING

The *National Electrical Code®* shows a separate table for each type of conduit. In order to keep *Ugly's Electrical References* in a compact and easy-to-use format, the following tables are included:

Electrical Metallic Tubing (EMT), Electrical Nonmetallic Tubing (ENT), PVC 40, PVC 80, Rigid Metal Conduit, Flexible Metal Conduit, and Liquidtight Flexible Metal Conduit.

When other types of conduit are used, refer to *NEC* Informative Annex C or use method shown below to figure conduit size.

Example #1 - All same wire size and type insulation.

10 – #12 RHH in Intermediate Metal Conduit.
*Go to the RHH Conductor Square Inch Area Table. (*Ugly's *page 98)*
#12 RHH = 0.0353 sq. in.   10 x 0.0353 sq. in. = 0.353 sq. in.
*Go to Intermediate Metal Conduit Square Inch Area Table. (*Ugly's *page 101)*
Use "Over 2 Wires 40%" column.
¾-inch conduit = 0.235 sq. in. (less than 0.353, so it's too small).
1-inch conduit = 0.384 sq. in. (greater than 0.353, so it's correct size).

Example #2 - Different wire sizes or types insulation.

10 – #12 RHH and 10 – #10 THHN in Liquidtight Nonmetallic Conduit (LFNC-B).
*Go to the RHH Conductor Square Inch Area Table. (*Ugly's *page 98)*
#12 RHH = 0.0353 sq. in.   10 x 0.0353 sq. in. = 0.353 sq. in.
*Go to the THHN Conductor Square Inch Area Table. (*Ugly's *page 98)*
#10 THHN = 0.0211 sq. in.   10 x 0.0211 sq. in. = 0.211 sq. in.
0.353 sq. in. + 0.211 sq. in. = 0.564 sq. in.
*Go to Liquidtight Flexible Nonmetallic Conduit (LFNC-B) Square Inch Table. (*Ugly's *page 102)*
Use "Over 2 Wires 40%" column.
1-inch conduit = 0.349 sq. in. (less than 0.564, so it's too small).
1¼-inch conduit = 0.611 sq. in. (greater than 0.564, so it's the correct size).

NOTE 1:* All conductors must be counted including grounding conductors for fill percentage.
NOTE 2: When all conductors are same type and size, decimals 0.8 and larger must be rounded up.
*NOTE 3**:* These are minimum size calculations, under certain conditions jamming can occur and the next size conduit must be used.
*NOTE 4***:* CAUTION - When over three current carrying conductors are used in the same circuit, conductor ampacity must be lower (adjusted).

* See Appendix C and Chapter 9 2014 *NEC* for complete tables and examples.
** See Chapter 9 Table 1 and Notes to Tables 1–9, 2014 *NEC*.
*** See 2014 *NEC* 310.15 for adjustment factors for temperature and number of conductors.

MAXIMUM NUMBER OF CONDUCTORS IN ELECTRICAL METALLIC TUBING

Type Letters	Cond. Size AWG/kcmil	Trade Sizes In Inches									
		½	¾	1	1¼	1½	2	2½	3	3½	4
RHH, RHW, RHW-2	14	4	7	11	20	27	46	80	120	157	201
	12	3	6	9	17	23	38	66	100	131	167
	10	2	5	8	13	18	30	53	81	105	135
	8	1	2	4	7	9	16	28	42	55	70
	6	1	1	3	5	8	13	22	34	44	56
	4	1	1	2	4	6	10	17	26	34	44
	3	1	1	1	4	5	9	15	23	30	38
	2	1	1	1	3	4	7	13	20	26	33
	1	0	1	1	1	3	5	9	13	17	22
	1/0	0	1	1	1	2	4	7	11	15	19
	2/0	0	1	1	1	2	4	6	10	13	17
	3/0	0	0	1	1	1	3	5	8	11	14
	4/0	0	0	1	1	1	3	5	7	9	12
	250	0	0	0	1	1	1	3	5	7	9
	300	0	0	0	1	1	1	3	5	6	8
	350	0	0	0	1	1	1	3	4	6	7
	400	0	0	0	1	1	1	2	4	5	7
	500	0	0	0	0	1	1	2	3	4	6
	600	0	0	0	0	1	1	1	3	4	5
	700	0	0	0	0	0	1	1	2	3	4
	750	0	0	0	0	0	1	1	2	3	4
TW	14	8	15	25	43	58	96	168	254	332	424
	12	6	11	19	33	45	74	129	195	255	326
	10	5	8	14	24	33	55	96	145	190	243
	8	2	5	8	13	18	30	53	81	105	135
RHH*, RHW*, RHW-2*, THHW, THW, THW-2	14	6	10	16	28	39	64	112	169	221	282
RHH*, RHW*, RHW-2*, THHW, THW	12	4	8	13	23	31	51	90	136	177	227
	10	3	6	10	18	24	40	70	106	138	177
RHH*, RHW*, RHW-2*, THHW, THW, THW-2	8	1	4	6	10	14	24	42	63	83	106
RHH*, RHW*, RHW-2*, TW, THW, THHW. THW-2	6	1	3	4	8	11	18	32	48	63	81
	4	1	1	3	6	8	13	24	36	47	60
	3	1	1	3	5	7	12	20	31	40	52
	2	1	1	2	4	6	10	17	26	34	44
	1	1	1	1	3	4	7	12	18	24	31
	1/0	0	1	1	2	3	6	10	16	20	26
	2/0	0	1	1	1	3	5	9	13	17	22
	3/0	0	1	1	1	2	4	7	11	15	19
	4/0	0	0	1	1	1	3	6	9	12	16
	250	0	0	1	1	1	3	5	7	10	13
	300	0	0	1	1	1	2	4	6	8	11
	350	0	0	0	1	1	1	4	6	7	10
	400	0	0	0	1	1	1	3	5	7	9
	500	0	0	0	1	1	1	3	4	6	7
	600	0	0	0	1	1	1	2	3	4	6
	700	0	0	0	0	1	1	1	3	4	5
	750	0	0	0	0	1	1	1	3	4	5
THHN, THWN, THWN-2	14	12	22	35	61	84	138	241	364	476	608
	12	9	16	26	45	61	101	176	266	347	443
	10	5	10	16	28	38	63	111	167	219	279
	8	3	6	9	16	22	36	64	96	126	161
	6	2	4	7	12	16	26	46	69	91	116
	4	1	2	4	7	10	16	28	43	56	71
	3	1	1	3	6	8	13	24	36	47	60
	2	1	1	3	5	7	11	20	30	40	51
	1	1	1	1	4	5	8	15	22	29	37

MAXIMUM NUMBER OF CONDUCTORS IN ELECTRICAL METALLIC TUBING

Type Letters	Cond. Size AWG/kcmil	Trade Sizes In Inches ½	¾	1	1¼	1½	2	2½	3	3½	4
THHN, THWN, THWN-2	1/0	1	1	1	3	4	7	12	19	25	32
	2/0	0	1	1	2	3	6	10	16	20	26
	3/0	0	1	1	1	3	5	8	13	17	22
	4/0	0	1	1	1	2	4	7	11	14	18
	250	0	0	1	1	1	3	6	9	11	15
	300	0	0	1	1	1	3	5	7	10	13
	350	0	0	1	1	1	2	4	6	9	11
	400	0	0	0	1	1	1	4	6	8	10
	500	0	0	0	1	1	1	3	5	6	8
	600	0	0	0	1	1	1	2	4	5	7
	700	0	0	0	1	1	1	2	3	4	6
	750	0	0	0	0	1	1	1	3	4	5
FEP, FEPB, PFA, PFAH, TFE	14	12	21	34	60	81	134	234	354	462	590
	12	9	15	25	43	59	98	171	258	337	430
	10	6	11	18	31	42	70	122	185	241	309
	8	3	6	10	18	24	40	70	106	138	177
	6	2	4	7	12	17	28	50	75	98	126
	4	1	3	5	9	12	20	35	53	69	88
	3	1	2	4	7	10	16	29	44	57	73
	2	1	1	3	6	8	13	24	36	47	60
PFA, PFAH, TFE	1	1	1	2	4	6	9	16	25	33	42
PFA, PFAH, TFE, Z	1/0	1	1	1	3	5	8	14	21	27	35
	2/0	0	1	1	3	4	6	11	17	22	29
	3/0	0	1	1	2	3	5	9	14	18	24
	4/0	0	1	1	1	2	4	8	11	15	19
Z	14	14	25	41	72	98	161	282	426	556	711
	12	10	18	29	51	69	114	200	302	394	504
	10	6	11	18	31	42	70	122	185	241	309
	8	4	7	11	20	27	44	77	117	153	195
	6	3	5	8	14	19	31	54	82	107	137
	4	1	3	5	9	13	21	37	56	74	94
	3	1	2	4	7	9	15	27	41	54	69
	2	1	1	3	6	8	13	22	34	45	57
	1	1	1	2	4	6	10	18	28	36	46
XHH, XHHW, XHHW-2, ZW	14	8	15	25	43	58	96	168	254	332	424
	12	6	11	19	33	45	74	129	195	255	326
	10	5	8	14	24	33	55	96	145	190	243
	8	2	5	8	13	18	30	53	81	105	135
	6	1	3	6	10	14	22	39	60	78	100
	4	1	2	4	7	10	16	28	43	56	72
	3	1	1	3	6	8	14	24	36	48	61
	2	1	1	3	5	7	11	20	31	40	51
XHH, XHHW, XHHW-2	1	1	1	1	4	5	8	15	23	30	38
	1/0	1	1	1	3	4	7	13	19	25	32
	2/0	0	1	1	2	3	6	10	16	21	27
	3/0	0	1	1	1	3	5	9	13	17	22
	4/0	0	1	1	1	2	4	7	11	14	18
	250	0	0	1	1	1	3	6	9	12	15
	300	0	0	1	1	1	3	5	8	10	13
	350	0	0	1	1	1	2	4	7	9	11
	400	0	0	0	1	1	1	4	6	8	10
	500	0	0	0	1	1	1	3	5	6	8
	600	0	0	0	1	1	1	2	4	5	6
	700	0	0	0	0	1	1	2	3	4	6
	750	0	0	0	0	1	1	1	3	4	5

*Types RHH, RHW, and RHW-2 without outer covering.

See *Ugly's* page 137 for Trade Size / Metric Designator conversion.

Source: NFPA 70, *National Electrical Code*®, NFPA, Quincy, MA, 2013, Table C.1, as modified.

MAXIMUM NUMBER OF CONDUCTORS IN NONMETALLIC TUBING

Type Letters	Cond. Size AWG/kcmil	Trade Sizes In Inches ½	¾	1	1¼	1½	2
RHH, RHW, RHW-2	14	3	6	10	19	26	43
	12	2	5	9	16	22	36
	10	1	4	7	13	17	29
	8	1	1	3	6	9	15
	6	1	1	3	5	7	12
	4	1	1	2	4	6	9
	3	1	1	1	3	5	8
	2	0	1	1	3	4	7
	1	0	1	1	1	3	5
	1/0	0	0	1	1	2	4
	2/0	0	0	1	1	1	3
	3/0	0	0	1	1	1	3
	4/0	0	0	1	1	1	2
	250	0	0	0	1	1	1
	300	0	0	0	1	1	1
	350	0	0	0	1	1	1
	400	0	0	0	1	1	1
	500	0	0	0	0	1	1
	600	0	0	0	0	1	1
	700	0	0	0	0	0	1
	750	0	0	0	0	0	1
TW	14	7	13	22	40	55	92
	12	5	10	17	31	42	71
	10	4	7	13	23	32	52
	8	1	4	7	13	17	29
RHH*, RHW*, RHW-2*, THHW, THW, THW-2	14	4	8	15	27	37	61
RHH*, RHW*, RHW-2*, THHW, THW	12	3	7	12	21	29	49
	10	3	5	9	17	23	38
RHH*, RHW*, RHW-2*, THHW, THW, THW-2	8	1	3	5	10	14	23
RHH*, RHW*, RHW-2*, TW, THW, THHW, THW-2	6	1	2	4	7	10	17
	4	1	1	3	5	8	13
	3	1	1	2	5	7	11
	2	1	1	2	4	6	9
	1	0	1	1	3	4	6
	1/0	0	1	1	2	3	5
	2/0	0	1	1	1	3	5
	3/0	0	0	1	1	2	4
	4/0	0	0	1	1	1	3
	250	0	0	1	1	1	2
	300	0	0	0	1	1	2
	350	0	0	0	1	1	1
	400	0	0	0	1	1	1
	500	0	0	0	1	1	1
	600	0	0	0	0	1	1
	700	0	0	0	0	1	1
	750	0	0	0	0	1	1
THHN, THWN, THWN-2	14	10	18	32	58	80	132
	12	7	13	23	42	58	96
	10	4	8	15	26	36	60
	8	2	5	8	15	21	35
	6	1	3	6	11	15	25
	4	1	1	4	7	9	15
	3	1	1	3	5	8	13
	2	1	1	2	5	6	11
	1	1	1	1	3	5	8

MAXIMUM NUMBER OF CONDUCTORS IN NONMETALLIC TUBING

Type Letters	Cond. Size AWG/kcmil	Trade Sizes In Inches ½	¾	1	1¼	1½	2
THHN, THWN, THWN-2	1/0	0	1	1	3	4	7
	2/0	0	1	1	2	3	5
	3/0	0	1	1	1	3	4
	4/0	0	0	1	1	2	4
	250	0	0	1	1	1	3
	300	0	0	1	1	1	2
	350	0	0	0	1	1	2
	400	0	0	0	1	1	1
	500	0	0	0	1	1	1
	600	0	0	0	1	1	1
	700	0	0	0	0	1	1
	750	0	0	0	0	1	1
FEP, FEPB, PFA, PFAH, TFE	14	10	18	31	56	77	128
	12	7	13	23	41	56	93
	10	5	9	16	29	40	67
	8	3	5	9	17	23	38
	6	1	4	6	12	16	27
	4	1	2	4	8	11	19
	3	1	1	4	7	9	16
	2	1	1	3	5	8	13
PFA, PFAH, TFE	1	1	1	1	4	5	9
PFA, PFAH, TFE, Z	1/0	0	1	1	3	4	7
	2/0	0	1	1	2	4	6
	3/0	0	1	1	1	3	5
	4/0	0	1	1	1	2	4
Z	14	12	22	38	68	93	154
	12	8	15	27	48	66	109
	10	5	9	16	29	40	67
	8	3	6	10	18	25	42
	6	1	4	7	13	18	30
	4	1	3	5	9	12	20
	3	1	1	3	6	9	15
	2	1	1	3	5	7	12
	1	1	1	2	4	6	10
XHH, XHHW, XHHW-2, ZW	14	7	13	22	40	55	92
	12	5	10	17	31	42	71
	10	4	7	13	23	32	52
	8	1	4	7	13	17	29
	6	1	3	5	9	13	21
	4	1	1	4	7	9	15
	3	1	1	3	6	8	13
	2	1	1	2	5	6	11
XHH, XHHW, XHHW-2	1	1	1	1	3	5	8
	1/0	0	1	1	3	4	7
	2/0	0	1	1	2	3	6
	3/0	0	1	1	1	3	5
	4/0	0	0	1	1	2	4
	250	0	0	1	1	1	3
	300	0	0	1	1	1	3
	350	0	0	1	1	1	2
	400	0	0	0	1	1	1
	500	0	0	0	1	1	1
	600	0	0	0	1	1	1
	700	0	0	0	0	1	1
	750	0	0	0	0	1	1

Types RHH, RHW, and RHW-2 without outer covering.

See *Ugly's* page 137 for Trade Size / Metric Designator conversion.

Source: NFPA 70, *National Electrical Code*®, NFPA, Quincy, MA, 2013, Table C.2, as modified.

MAXIMUM NUMBER OF CONDUCTORS IN RIGID PVC CONDUIT, SCHEDULE 40

Type Letters	Cond. Size AWG/kcmil	Trade Sizes In Inches ½	¾	1	1¼	1½	2	2½	3	3½	4	5	6
RHH, RHW, RHW-2	14	4	7	11	20	27	45	64	99	133	171	269	390
	12	3	5	9	16	22	37	53	82	110	142	224	323
	10	2	4	7	13	18	30	43	66	89	115	181	261
	8	1	2	4	7	9	15	22	35	46	60	94	137
	6	1	1	3	5	7	12	18	28	37	48	76	109
	4	1	1	2	4	6	10	14	22	29	37	59	85
	3	1	1	1	4	5	8	12	19	25	33	52	75
	2	1	1	1	3	4	7	10	16	22	28	45	65
	1	0	1	1	1	3	5	7	11	14	19	29	43
	1/0	0	1	1	1	2	4	6	9	13	16	26	37
	2/0	0	0	1	1	1	3	5	8	11	14	22	32
	3/0	0	0	1	1	1	3	4	7	9	12	19	28
	4/0	0	0	1	1	1	2	4	6	8	10	16	24
	250	0	0	0	1	1	1	3	4	6	8	12	18
	300	0	0	0	1	1	1	2	4	5	7	11	16
	350	0	0	0	1	1	1	2	3	5	6	10	14
	400	0	0	0	1	1	1	1	3	4	6	9	13
	500	0	0	0	0	1	1	1	3	4	5	8	11
	600	0	0	0	0	1	1	1	2	3	4	6	9
	700	0	0	0	0	0	1	1	1	3	3	6	8
	750	0	0	0	0	0	1	1	1	2	3	5	8
TW	14	8	14	24	42	57	94	135	209	280	361	568	822
	12	6	11	18	32	44	72	103	160	215	277	436	631
	10	4	8	13	24	32	54	77	119	160	206	325	470
	8	2	4	7	13	18	30	43	66	89	115	181	261
RHH*, RHW*, RHW-2*, THHW, THW, THW-2	14	5	9	16	28	38	63	90	139	186	240	378	546
RHH*, RHW*, RHW-2*, THHW, THW	12	4	8	12	22	30	50	72	112	150	193	304	43
	10	3	6	10	17	24	39	56	87	117	150	237	34
RHH*, RHW*, RHW-2*, THHW, THW, THW-2	8	1	3	6	10	14	23	33	52	70	90	142	20
RHH*, RHW*, RHW-2*, TW, THW, THHW, THW-2	6	1	2	4	8	11	18	26	40	53	69	109	15
	4	1	1	3	6	8	13	19	30	40	51	81	11
	3	1	1	3	5	7	11	16	25	34	44	69	10
	2	1	1	2	4	6	10	14	22	29	37	59	8
	1	0	1	1	3	4	7	10	15	20	26	41	6
	1/0	0	1	1	2	3	6	8	13	17	22	35	5
	2/0	0	1	1	1	3	5	7	11	15	19	30	4
	3/0	0	1	1	1	2	4	6	9	12	16	25	3
	4/0	0	0	1	1	1	3	5	8	10	13	21	3
	250	0	0	1	1	1	3	4	6	8	11	17	2
	300	0	0	1	1	1	2	3	5	7	9	15	2
	350	0	0	0	1	1	1	3	5	6	8	13	1
	400	0	0	0	1	1	1	3	4	6	7	12	1
	500	0	0	0	1	1	1	2	3	5	6	10	1
	600	0	0	0	0	1	1	1	3	4	5	8	1
	700	0	0	0	0	1	1	1	2	3	4	7	1
	750	0	0	0	0	1	1	1	2	3	4	6	1
THHN, THWN, THWN-2	14	11	21	34	60	82	135	193	299	401	517	815	117
	12	8	15	25	43	59	99	141	218	293	377	594	85
	10	5	9	15	27	37	62	89	137	184	238	374	54
	8	3	5	9	16	21	36	51	79	106	137	216	31
	6	1	4	6	11	15	26	37	57	77	99	156	22
	4	1	2	4	7	9	16	22	35	47	61	96	13
	3	1	1	3	6	8	13	19	30	40	51	81	11
	2	1	1	3	5	7	11	16	25	33	43	68	9
	1	1	1	1	3	5	8	12	18	25	32	50	7

MAXIMUM NUMBER OF CONDUCTORS IN RIGID PVC CONDUIT, SCHEDULE 40

Type Letters	Cond. Size AWG/kcmil	Trade Sizes In Inches ½	¾	1	1¼	1½	2	2½	3	3½	4	5	6
THHN, THWN, THWN-2	1/0	1	1	1	3	4	7	10	15	21	27	42	61
	2/0	0	1	1	2	3	6	8	13	17	22	35	51
	3/0	0	1	1	1	3	5	7	11	14	18	29	42
	4/0	0	1	1	1	2	4	6	9	12	15	24	35
	250	0	0	1	1	1	3	4	7	10	12	20	28
	300	0	0	1	1	1	3	4	6	8	11	17	24
	350	0	0	1	1	1	2	3	5	7	9	15	21
	400	0	0	0	1	1	1	3	5	6	8	13	19
	500	0	0	0	1	1	1	2	4	5	7	11	16
	600	0	0	0	1	1	1	1	3	4	5	9	13
	700	0	0	0	0	1	1	1	3	4	5	8	11
	750	0	0	0	0	1	1	1	2	3	4	7	11
FEP, FEPB, PFA, PFAH, TFE	14	11	20	33	58	79	131	188	290	389	502	790	1142
	12	8	15	24	42	58	96	137	212	284	366	577	834
	10	6	10	17	30	41	69	98	152	204	263	414	598
	8	3	6	10	17	24	39	56	87	117	150	237	343
	6	2	4	7	12	17	28	40	62	83	107	169	244
	4	1	3	5	8	12	19	28	43	58	75	118	170
	3	1	2	4	7	10	16	23	36	48	62	98	142
	2	1	1	3	6	8	13	19	30	40	51	81	117
PFA, PFAH, TFE	1	1	1	2	4	5	9	13	20	28	36	56	81
PFA, PFAH, TFE, Z	1/0	1	1	1	3	4	8	11	17	23	30	47	68
	2/0	0	1	1	3	4	6	9	14	19	24	39	56
	3/0	0	1	1	2	3	5	7	12	16	20	32	46
	4/0	0	1	1	1	2	4	6	9	13	16	26	38
Z	14	13	24	40	70	95	158	226	350	469	605	952	1376
	12	9	17	28	49	68	112	160	248	333	429	675	976
	10	6	10	17	30	41	69	98	152	204	263	414	598
	8	3	6	11	19	26	43	62	96	129	166	261	378
	6	2	4	7	13	18	30	43	67	90	116	184	265
	4	1	3	5	9	12	21	30	46	62	80	126	183
	3	1	2	4	6	9	15	22	34	45	58	92	133
	2	1	1	3	5	7	12	18	28	38	49	77	111
	1	1	1	2	4	6	10	14	23	30	39	62	90
XHH, XHHW, XHHW-2, ZW	14	8	14	24	42	57	94	135	209	280	361	568	822
	12	6	11	18	32	44	72	103	160	215	277	436	631
	10	4	8	13	24	32	54	77	119	160	206	325	470
	8	2	4	7	13	18	30	43	66	89	115	181	261
	6	1	3	5	10	13	22	32	49	66	85	134	193
	4	1	2	4	7	9	16	23	35	48	61	97	140
	3	1	1	3	6	8	13	19	30	40	52	82	118
	2	1	1	3	5	7	11	16	25	34	44	69	99
XHH, XHHW, XHHW-2	1	1	1	1	3	5	8	12	19	25	32	51	74
	1/0	1	1	1	3	4	7	10	16	21	27	43	62
	2/0	0	1	1	2	3	6	8	13	17	23	36	52
	3/0	0	1	1	1	3	5	7	11	14	19	30	43
	4/0	0	1	1	1	2	4	6	9	12	15	24	35
	250	0	0	1	1	1	3	5	7	10	13	20	29
	300	0	0	1	1	1	3	4	6	8	11	17	25
	350	0	0	1	1	1	2	3	6	7	9	15	22
	400	0	0	0	1	1	1	3	5	6	8	13	19
	500	0	0	0	1	1	1	2	4	5	7	11	16
	600	0	0	0	1	1	1	1	3	4	5	9	13
	700	0	0	0	0	1	1	1	3	4	5	8	11
	750	0	0	0	0	1	1	1	2	3	4	7	11

Types RHH, RHW, and RHW-2 without outer covering.

ee *Ugly's* page 137 for Trade Size / Metric Designator conversion.

ource: NFPA 70, *National Electrical Code*®, NFPA, Quincy, MA, 2013, Table C.10, as modified.

MAXIMUM NUMBER OF CONDUCTORS IN RIGID PVC CONDUIT, SCHEDULE 80

Type Letters	Cond. Size AWG/kcmil	Trade Sizes In Inches											
		½	¾	1	1¼	1½	2	2½	3	3½	4	5	6
RHH, RHW, RHW-2	14	3	5	9	17	23	39	56	88	118	153	243	349
	12	2	4	7	14	19	32	46	73	98	127	202	290
	10	1	3	6	11	15	26	37	59	79	103	163	234
	8	1	1	3	6	8	13	19	31	41	54	85	122
	6	1	1	2	4	6	11	16	24	33	43	68	98
	4	1	1	1	3	5	8	12	19	26	33	53	77
	3	0	1	1	3	4	7	11	17	23	29	47	67
	2	0	1	1	3	4	6	9	14	20	25	41	58
	1	0	1	1	1	2	4	6	9	13	17	27	38
	1/0	0	0	1	1	1	3	5	8	11	15	23	33
	2/0	0	0	1	1	1	3	4	7	10	13	20	29
	3/0	0	0	1	1	1	3	4	6	8	11	17	25
	4/0	0	0	0	1	1	2	3	5	7	9	15	21
	250	0	0	0	1	1	1	2	4	5	7	11	16
	300	0	0	0	1	1	1	2	3	5	6	10	14
	350	0	0	0	1	1	1	1	3	4	5	9	13
	400	0	0	0	0	1	1	1	3	4	5	8	12
	500	0	0	0	0	1	1	1	2	3	4	7	10
	600	0	0	0	0	0	1	1	1	3	3	6	8
	700	0	0	0	0	0	1	1	1	2	3	5	7
	750	0	0	0	0	0	1	1	1	2	3	5	7
TW	14	6	11	20	35	49	82	118	185	250	324	514	736
	12	5	9	15	27	38	63	91	142	192	248	394	565
	10	3	6	11	20	28	47	67	106	143	185	294	421
	8	1	3	6	11	15	26	37	59	79	103	163	234
RHH*, RHW*, RHW-2*, THHW, THW, THW-2	14	4	8	13	23	32	55	79	123	166	215	341	490
RHH*, RHW*, RHW-2*, THHW, THW	12	3	6	10	19	26	44	63	99	133	173	274	394
	10	2	5	8	15	20	34	49	77	104	135	214	307
RHH*, RHW*, RHW-2*, THHW, THW, THW-2	8	1	3	5	9	12	20	29	46	62	81	128	184
RHH*, RHW*, RHW-2*, TW, THW, THHW, THW-2	6	1	1	3	7	9	16	22	35	48	62	98	141
	4	1	1	3	5	7	12	17	26	35	46	73	105
	3	1	1	2	4	6	10	14	22	30	39	63	90
	2	1	1	1	3	5	8	12	19	26	33	53	77
	1	0	1	1	2	3	6	8	13	18	23	37	54
	1/0	0	1	1	1	3	5	7	11	15	20	32	46
	2/0	0	1	1	1	2	4	6	10	13	17	27	39
	3/0	0	0	1	1	1	3	5	8	11	14	23	33
	4/0	0	0	1	1	1	3	4	7	9	12	19	27
	250	0	0	0	1	1	2	3	5	7	9	15	22
	300	0	0	0	1	1	1	3	5	6	8	13	19
	350	0	0	0	1	1	1	2	4	6	7	12	1[illegible]
	400	0	0	0	1	1	1	2	4	5	7	10	1[illegible]
	500	0	0	0	1	1	1	1	3	4	5	9	1[illegible]
	600	0	0	0	0	1	1	1	2	3	4	7	1[illegible]
	700	0	0	0	0	1	1	1	2	3	4	6	[illegible]
	750	0	0	0	0	0	1	1	1	3	4	6	[illegible]
THHN, THWN, THWN-2	14	9	17	28	51	70	118	170	265	358	464	736	105[illegible]
	12	6	12	20	37	51	86	124	193	261	338	537	77[illegible]
	10	4	7	13	23	32	54	78	122	164	213	338	48[illegible]
	8	2	4	7	13	18	31	45	70	95	123	195	27[illegible]
	6	1	3	5	9	13	22	32	51	68	89	141	20[illegible]
	4	1	1	3	6	8	14	20	31	42	54	86	12[illegible]
	3	1	1	3	5	7	12	17	26	35	46	73	10[illegible]
	2	1	1	2	4	6	10	14	22	30	39	61	8[illegible]
	1	0	1	1	3	4	7	10	16	22	29	45	6[illegible]

MAXIMUM NUMBER OF CONDUCTORS IN RIGID PVC CONDUIT, SCHEDULE 80

Type Letters	Cond. Size AWG/kcmil	Trade Sizes In Inches ½	¾	1	1¼	1½	2	2½	3	3½	4	5	6
THHN, THWN, THWN-2	1/0	0	1	1	2	3	6	9	14	18	24	38	55
	2/0	0	1	1	1	3	5	7	11	15	20	32	46
	3/0	0	1	1	1	2	4	6	9	13	17	26	38
	4/0	0	0	1	1	1	3	5	8	10	14	22	31
	250	0	0	1	1	1	3	4	6	8	11	18	25
	300	0	0	0	1	1	2	3	5	7	9	15	22
	350	0	0	0	1	1	1	3	5	6	8	13	19
	400	0	0	0	1	1	1	3	4	6	7	12	17
	500	0	0	0	1	1	1	2	3	5	6	10	14
	600	0	0	0	0	1	1	1	3	4	5	8	12
	700	0	0	0	0	1	1	1	2	3	4	7	10
	750	0	0	0	0	1	1	1	2	3	4	7	9
FEP, FEPB, PFA, PFAH, TFE	14	8	16	27	49	68	115	164	257	347	450	714	1024
	12	6	12	20	36	50	84	120	188	253	328	521	747
	10	4	8	14	26	36	60	86	135	182	235	374	536
	8	2	5	8	15	20	34	49	77	104	135	214	307
	6	1	3	6	10	14	24	35	55	74	96	152	218
	4	1	2	4	7	10	17	24	38	52	67	106	153
	3	1	1	3	6	8	14	20	32	43	56	89	127
	2	1	1	3	5	7	12	17	26	35	46	73	105
PFA, PFAH, TFE	1	1	1	1	3	5	8	11	18	25	32	51	73
PFA, PFAH, TFE, Z	1/0	0	1	1	3	4	7	10	15	20	27	42	61
	2/0	0	1	1	2	3	5	8	12	17	22	35	50
	3/0	0	1	1	1	2	4	6	10	14	18	29	41
	4/0	0	0	1	1	1	4	5	8	11	15	24	34
Z	14	10	19	33	59	82	138	198	310	418	542	860	1233
	12	7	14	23	42	58	98	141	220	297	385	610	875
	10	4	8	14	26	36	60	86	135	182	235	374	536
	8	3	5	9	16	22	38	54	85	115	149	236	339
	6	2	4	6	11	16	26	38	60	81	104	166	238
	4	1	2	4	8	11	18	26	41	55	72	114	164
	3	1	2	3	5	8	13	19	30	40	52	83	119
	2	1	1	2	5	6	11	16	25	33	43	69	99
	1	0	1	2	4	5	9	13	20	27	35	56	80
XHH, XHHW, XHHW-2, ZW	14	6	11	20	35	49	82	118	185	250	324	514	736
	12	5	9	15	27	38	63	91	142	192	248	394	565
	10	3	6	11	20	28	47	67	106	143	185	294	421
	8	1	3	6	11	15	26	37	59	79	103	163	234
	6	1	2	4	8	11	19	28	43	59	76	121	173
	4	1	1	3	6	8	14	20	31	42	55	87	125
	3	1	1	3	5	7	12	17	26	36	47	74	106
	2	1	1	2	4	6	10	14	22	30	39	62	89
XHH, XHHW, XHHW-2	1	0	1	1	3	4	7	10	16	22	29	46	66
	1/0	0	1	1	2	3	6	9	14	19	24	39	56
	2/0	0	1	1	1	3	5	7	11	16	20	32	46
	3/0	0	1	1	1	2	4	6	9	13	17	27	38
	4/0	0	0	1	1	1	3	5	8	11	14	22	32
	250	0	0	1	1	1	3	4	6	9	11	18	26
	300	0	0	1	1	1	2	3	5	7	10	15	22
	350	0	0	0	1	1	1	3	5	6	8	14	20
	400	0	0	0	1	1	1	0	4	6	7	12	17
	500	0	0	0	1	1	1	2	3	5	6	10	14
	600	0	0	0	0	1	1	1	3	4	5	8	11
	700	0	0	0	0	1	1	1	2	3	4	7	10
	750	0	0	0	0	1	1	1	2	3	4	6	9

*Types RHH, RHW, and RHW-2 without outer covering.

See *Ugly's* page 137 for Trade Size / Metric Designator conversion.

Source: NFPA 70, *National Electrical Code®*, NFPA, Quincy, MA, 2013, Table C.9, as modified.

MAXIMUM NUMBER OF CONDUCTORS IN RIGID METAL CONDUIT

Type Letters	Cond. Size AWG/kcmil	Trade Sizes In Inches ½	¾	1	1¼	1½	2	2½	3	3½	4	5	6
RHH, RHW, RHW-2	14	4	7	12	21	28	46	66	102	136	176	276	398
	12	3	6	10	17	23	38	55	85	113	146	229	330
	10	3	5	8	14	19	31	44	68	91	118	185	267
	8	1	2	4	7	10	16	23	36	48	61	97	139
	6	1	1	3	6	8	13	18	29	38	49	77	112
	4	1	1	2	4	6	10	14	22	30	38	60	87
	3	1	1	2	4	5	9	12	19	26	34	53	76
	2	1	1	1	3	4	7	11	17	23	29	46	66
	1	0	1	1	1	3	5	7	11	15	19	30	44
	1/0	0	1	1	1	2	4	6	10	13	17	26	38
	2/0	0	1	1	1	2	4	5	8	11	14	23	33
	3/0	0	0	1	1	1	3	4	7	10	12	20	28
	4/0	0	0	1	1	1	3	4	6	8	11	17	24
	250	0	0	0	1	1	1	3	4	6	8	13	18
	300	0	0	0	1	1	1	2	4	5	7	11	16
	350	0	0	0	1	1	1	2	4	5	6	10	15
	400	0	0	0	1	1	1	1	3	4	6	9	13
	500	0	0	0	1	1	1	1	3	4	5	8	11
	600	0	0	0	0	1	1	1	2	3	4	6	9
	700	0	0	0	0	1	1	1	1	3	4	6	8
	750	0	0	0	0	0	1	1	1	3	3	5	8
TW	14	9	15	25	44	59	98	140	216	288	370	581	839
	12	7	12	19	33	45	75	107	165	221	284	446	644
	10	5	9	14	25	34	56	80	123	164	212	332	480
	8	3	5	8	14	19	31	44	68	91	118	185	267
RHH*, RHW*, RHW-2*, THHW, THW, THW-2	14	6	10	17	29	39	65	93	143	191	246	387	558
RHH*, RHW*, RHW-2*, THHW, THW	12	5	8	13	23	32	52	75	115	154	198	311	448
	10	3	6	10	18	25	41	58	90	120	154	242	350
RHH*, RHW*, RHW-2*, THHW, THW, THW-2	8	1	4	6	11	15	24	35	54	72	92	145	209
RHH*, RHW*, RHW-2*, TW, THW, THHW, THW-2	6	1	3	5	8	11	18	27	41	55	71	111	160
	4	1	1	3	6	8	14	20	31	41	53	83	120
	3	1	1	3	5	7	12	17	26	35	45	71	103
	2	1	1	2	4	6	10	14	22	30	38	60	87
	1	1	1	1	3	4	7	10	15	21	27	42	61
	1/0	0	1	1	2	3	6	8	13	18	23	36	52
	2/0	0	1	1	2	3	5	7	11	15	19	31	44
	3/0	0	1	1	1	2	4	6	9	13	16	26	37
	4/0	0	0	1	1	1	3	5	8	10	14	21	31
	250	0	0	1	1	1	3	4	6	8	11	17	25
	300	0	0	1	1	1	2	3	5	7	9	15	22
	350	0	0	0	1	1	1	3	5	6	8	13	19
	400	0	0	0	1	1	1	3	4	6	7	12	17
	500	0	0	0	1	1	1	2	3	5	6	10	14
	600	0	0	0	1	1	1	1	3	4	5	8	12
	700	0	0	0	0	1	1	1	2	3	4	7	10
	750	0	0	0	0	1	1	1	2	3	4	7	10
THHN, THWN, THWN-2	14	13	22	36	63	85	140	200	309	412	531	833	1202
	12	9	16	26	46	62	102	146	225	301	387	608	877
	10	6	10	17	29	39	64	92	142	189	244	383	552
	8	3	6	9	16	22	37	53	82	109	140	221	318
	6	2	4	7	12	16	27	38	59	79	101	159	230
	4	1	2	4	7	10	16	23	36	48	62	98	141
	3	1	1	3	6	8	14	20	31	41	53	83	120
	2	1	1	3	5	7	11	17	26	34	44	70	100
	1	1	1	1	4	5	8	12	19	25	33	51	74

MAXIMUM NUMBER OF CONDUCTORS IN RIGID METAL CONDUIT

Type Letters	Cond. Size AWG/kcmil	Trade Sizes In Inches ½	¾	1	1¼	1½	2	2½	3	3½	4	5	6
THHN, THWN, THWN-2	1/0	1	1	1	3	4	7	10	16	21	27	43	63
	2/0	0	1	1	2	3	6	8	13	18	23	36	52
	3/0	0	1	1	1	3	5	7	11	15	19	30	43
	4/0	0	1	1	1	2	4	6	9	12	16	25	36
	250	0	0	1	1	1	3	5	7	10	13	20	29
	300	0	0	1	1	1	3	4	6	8	11	17	25
	350	0	0	1	1	1	2	3	5	7	10	15	22
	400	0	0	1	1	1	2	3	5	7	8	13	20
	500	0	0	0	1	1	1	2	4	5	7	11	16
	600	0	0	0	1	1	1	1	3	4	6	9	13
	700	0	0	0	1	1	1	1	3	4	5	8	11
	750	0	0	0	0	1	1	1	3	4	5	7	11
FEP, FEPB, PFA, PFAH, TFE	14	12	22	35	61	83	136	194	300	400	515	808	1166
	12	9	16	26	44	60	99	142	219	292	376	590	851
	10	6	11	18	32	43	71	102	157	209	269	423	610
	8	3	6	10	18	25	41	58	90	120	154	242	350
	6	2	4	7	13	17	29	41	64	85	110	172	249
	4	1	3	5	9	12	20	29	44	59	77	120	174
	3	1	2	4	7	10	17	24	37	50	64	100	145
	2	1	1	3	6	8	14	20	31	41	53	83	120
PFA, PFAH, TFE	1	1	1	2	4	6	9	14	21	28	37	57	83
PFA, PFAH, TFE, Z	1/0	1	1	1	3	5	8	11	18	24	30	48	69
	2/0	1	1	1	3	4	6	9	14	19	25	40	57
	3/0	0	1	1	2	3	5	8	12	16	21	33	47
	4/0	0	1	1	1	2	4	6	10	13	17	27	39
Z	14	15	26	42	73	100	164	234	361	482	621	974	1405
	12	10	18	30	52	71	116	166	256	342	440	691	997
	10	6	11	18	32	43	71	102	157	209	269	423	610
	8	4	7	11	20	27	45	64	99	132	170	267	386
	6	3	5	8	14	19	31	45	69	93	120	188	271
	4	1	3	5	9	13	22	31	48	64	82	129	186
	3	1	2	4	7	9	16	22	35	47	60	94	136
	2	1	1	3	6	8	13	19	29	39	50	78	113
	1	1	1	2	5	6	10	15	23	31	40	63	92
XHH, XHHW, XHHW-2, ZW	14	9	15	25	44	59	98	140	216	288	370	581	839
	12	7	12	19	33	45	75	107	165	221	284	446	644
	10	5	9	14	25	34	56	80	123	164	212	332	480
	8	3	5	8	14	19	31	44	68	91	118	185	267
	6	1	3	6	10	14	23	33	51	68	87	137	197
	4	1	2	4	7	10	16	24	37	49	63	99	143
	3	1	1	3	6	8	14	20	31	41	53	84	121
	2	1	1	3	5	7	12	17	26	35	45	70	101
	1	1	1	1	4	5	9	12	19	26	33	52	76
XHH, XHHW, XHHW-2	1/0	1	1	1	3	4	7	10	16	22	28	44	64
	2/0	0	1	1	2	3	6	9	13	18	23	37	53
	3/0	0	1	1	1	3	5	7	11	15	19	30	44
	4/0	0	1	1	1	2	4	6	9	12	16	25	36
	250	0	0	1	1	1	3	5	7	10	13	20	30
	300	0	0	1	1	1	3	4	6	9	11	18	25
	350	0	0	1	1	1	2	3	6	7	10	15	22
	400	0	0	1	1	1	2	3	5	7	9	14	20
	500	0	0	0	1	1	1	2	4	5	7	11	16
	600	0	0	0	1	1	1	1	3	4	6	9	13
	700	0	0	0	1	1	1	1	3	4	5	8	11
	750	0	0	0	0	1	1	1	3	4	5	7	11

*Types RHH, RHW, and RHW-2 without outer covering.

See *Ugly's* page 137 for Trade Size / Metric Designator conversion.

Source: NFPA 70, *National Electrical Code*®, NFPA, Quincy, MA, 2013, Table C.8, as modified.

MAXIMUM NUMBER OF CONDUCTORS IN FLEXIBLE METAL CONDUIT

Type Letters	Cond. Size AWG/kcmil	Trade Sizes In Inches ½	¾	1	1¼	1½	2	2½	3	3½	4
RHH, RHW, RHW-2	14	4	7	11	17	25	44	67	96	131	171
	12	3	6	9	14	21	37	55	80	109	142
	10	3	5	7	11	17	30	45	64	88	115
	8	1	2	4	6	9	15	23	34	46	60
	6	1	1	3	5	7	12	19	27	37	48
	4	1	1	2	4	5	10	14	21	29	37
	3	1	1	1	3	5	8	13	18	25	33
	2	1	1	1	3	4	7	11	16	22	28
	1	0	1	1	1	2	5	7	10	14	19
	1/0	0	1	1	1	2	4	6	9	12	16
	2/0	0	1	1	1	2	3	5	8	11	14
	3/0	0	0	1	1	1	3	5	7	9	12
	4/0	0	0	1	1	1	2	4	6	8	10
	250	0	0	0	1	1	1	3	4	6	8
	300	0	0	0	1	1	1	2	4	5	7
	350	0	0	0	1	1	1	2	3	5	6
	400	0	0	0	0	1	1	1	3	4	6
	500	0	0	0	0	1	1	1	3	4	5
	600	0	0	0	0	1	1	1	2	3	4
	700	0	0	0	0	0	1	1	1	3	3
	750	0	0	0	0	0	1	1	1	2	3
TW	14	9	15	23	36	53	94	141	203	277	361
	12	7	11	18	28	41	72	108	156	212	277
	10	5	8	13	21	30	54	81	116	158	207
	8	3	5	7	11	17	30	45	64	88	115
RHH*, RHW*, RHW-2*, THHW, THW, THW-2	14	6	10	15	24	35	62	94	135	184	240
RHH*, RHW*, RHW-2*, THHW, THW	12	5	8	12	19	28	50	75	108	148	193
	10	4	6	10	15	22	39	59	85	115	151
RHH*, RHW*, RHW-2*, THHW, THW, THW-2	8	1	4	6	9	13	23	35	51	69	90
RHH*, RHW*, RHW-2*, TW, THW, THHW, THW-2	6	1	3	4	7	10	18	27	39	53	69
	4	1	1	3	5	7	13	20	29	39	51
	3	1	1	3	4	6	11	17	25	34	44
	2	1	1	2	4	5	10	14	21	29	37
	1	1	1	1	2	4	7	10	15	20	26
	1/0	0	1	1	1	3	6	9	12	17	22
	2/0	0	1	1	1	3	5	7	10	14	19
	3/0	0	1	1	1	2	4	6	9	12	16
	4/0	0	0	1	1	1	3	5	7	10	13
	250	0	0	1	1	1	3	4	6	8	11
	300	0	0	1	1	1	2	3	5	7	9
	350	0	0	0	1	1	1	3	4	6	8
	400	0	0	0	1	1	1	3	4	6	7
	500	0	0	0	1	1	1	2	3	5	6
	600	0	0	0	0	1	1	1	3	4	5
	700	0	0	0	0	1	1	1	2	3	4
	750	0	0	0	0	1	1	1	2	3	4
THHN, THWN, THWN-2	14	13	22	33	52	76	134	202	291	396	518
	12	9	16	24	38	56	98	147	212	289	378
	10	6	10	15	24	35	62	93	134	182	238
	8	3	6	9	14	20	35	53	77	105	137
	6	2	4	6	10	14	25	38	55	76	99
	4	1	2	4	6	9	16	24	34	46	61
	3	1	1	3	5	7	13	20	29	39	51
	2	1	1	3	4	6	11	17	24	33	43
	1	1	1	1	3	4	8	12	18	24	32

MAXIMUM NUMBER OF CONDUCTORS IN FLEXIBLE METAL CONDUIT

Type Letters	Cond. Size AWG/kcmil	½	¾	1	1¼	1½	2	2½	3	3½	4
THHN, THWN, THWN-2	1/0	1	1	1	2	4	7	10	15	20	27
	2/0	0	1	1	1	3	6	9	12	17	22
	3/0	0	1	1	1	2	5	7	10	14	18
	4/0	0	1	1	1	1	4	6	8	12	15
	250	0	0	1	1	1	3	5	7	9	12
	300	0	0	1	1	1	3	4	6	8	11
	350	0	0	1	1	1	2	3	5	7	9
	400	0	0	0	1	1	1	3	5	6	8
	500	0	0	0	1	1	1	2	4	5	7
	600	0	0	0	0	1	1	1	3	4	5
	700	0	0	0	0	1	1	1	3	4	5
	750	0	0	0	0	1	1	1	2	3	4
FEP, FEPB, PFA, PFAH, TFE	14	12	21	32	51	74	130	196	282	385	502
	12	9	15	24	37	54	95	143	206	281	367
	10	6	11	17	26	39	68	103	148	201	263
	8	4	6	10	15	22	39	59	85	115	151
	6	2	4	7	11	16	28	42	60	82	107
	4	1	3	5	7	11	19	29	42	57	75
	3	1	2	4	6	9	16	24	35	48	62
	2	1	1	3	5	7	13	20	29	39	51
PFA, PFAH, TFE	1	1	1	2	3	5	9	14	20	27	36
PFA, PFAH, TFE, Z	1/0	1	1	1	3	4	8	11	17	23	30
	2/0	1	1	1	2	3	6	9	14	19	24
	3/0	0	1	1	1	3	5	8	11	15	20
	4/0	0	1	1	1	2	4	6	9	13	16
Z	14	15	25	39	61	89	157	236	340	463	605
	12	11	18	28	43	63	111	168	241	329	429
	10	6	11	17	26	39	68	103	148	201	263
	8	4	7	11	17	24	43	65	93	127	166
	6	3	5	7	12	17	30	45	65	89	117
	4	1	3	5	8	12	21	31	45	61	80
	3	1	2	4	6	8	15	23	33	45	58
	2	1	1	3	5	7	12	19	27	37	49
	1	1	1	2	4	6	10	15	22	30	39
XHH, XHHW, XHHW-2, ZW	14	9	15	23	36	53	94	141	203	277	361
	12	7	11	18	28	41	72	108	156	212	277
	10	5	8	13	21	30	54	81	116	158	207
	8	3	5	7	11	17	30	45	64	88	115
	6	1	3	5	8	12	22	33	48	65	85
	4	1	2	4	6	9	16	24	34	47	61
	3	1	1	3	5	7	13	20	29	40	52
	2	1	1	3	4	6	11	17	24	33	44
XHH, XHHW, XHHW-2	1	1	1	1	3	5	8	13	18	25	32
	1/0	1	1	1	2	4	7	10	15	21	27
	2/0	0	1	1	2	3	6	9	13	17	23
	3/0	0	1	1	1	3	5	7	10	14	19
	4/0	0	1	1	1	2	4	6	9	12	15
	250	0	0	1	1	1	3	5	7	10	13
	300	0	0	1	1	1	3	4	6	8	11
	350	0	0	1	1	1	2	4	5	7	9
	400	0	0	0	1	1	1	3	5	6	8
	500	0	0	0	1	1	1	3	4	5	7
	600	0	0	0	0	1	1	1	3	4	5
	700	0	0	0	0	1	1	1	3	4	5
	750	0	0	0	0	1	1	1	2	3	4

*Types RHH, RHW, and RHW-2 without outer covering.

See *Ugly's* page 137 for Trade Size / Metric Designator conversion.

Source: NFPA 70, *National Electrical Code®*, NFPA, Quincy, MA, 2013, Table C.3, as modified.

MAXIMUM NUMBER OF CONDUCTORS IN LIQUIDTIGHT FLEXIBLE METAL CONDUIT

Type Letters	Cond. Size AWG/kcmil	Trade Sizes In Inches									
		½	¾	1	1¼	1½	2	2½	3	3½	4
RHH, RHW, RHW-2	14	4	7	12	21	27	44	66	102	133	173
	12	3	6	10	17	22	36	55	84	110	144
	10	3	5	8	14	18	29	44	68	89	116
	8	1	2	4	7	9	15	23	36	46	61
	6	1	1	3	6	7	12	18	28	37	48
	4	1	1	2	4	6	9	14	22	29	38
	3	1	1	1	4	5	8	13	19	25	33
	2	1	1	1	3	4	7	11	17	22	29
	1	0	1	1	1	3	5	7	11	14	19
	1/0	0	1	1	1	2	4	6	10	13	16
	2/0	0	1	1	1	1	3	5	8	11	14
	3/0	0	0	1	1	1	3	4	7	9	12
	4/0	0	0	1	1	1	2	4	6	8	10
	250	0	0	0	1	1	1	3	4	6	8
	300	0	0	0	1	1	1	2	4	5	7
	350	0	0	0	1	1	1	2	3	5	6
	400	0	0	0	1	1	1	1	3	4	6
	500	0	0	0	1	1	1	1	3	4	5
	600	0	0	0	0	1	1	1	2	3	4
	700	0	0	0	0	0	1	1	1	3	3
	750	0	0	0	0	0	1	1	1	2	3
TW	14	9	15	25	44	57	93	140	215	280	365
	12	7	12	19	33	43	71	108	165	215	280
	10	5	9	14	25	32	53	80	123	160	209
	8	3	5	8	14	18	29	44	68	89	116
RHH*, RHW*, RHW-2*, THHW, THW, THW-2	14	6	10	16	29	38	62	93	143	186	243
RHH*, RHW*, RHW-2*, THHW, THW	12	5	8	13	23	30	50	75	115	149	195
	10	3	6	10	18	23	39	58	89	117	152
RHH*, RHW*, RHW-2*, THHW, THW, THW-2	8	1	4	6	11	14	23	35	53	70	91
RHH*, RHW*, RHW-2*, TW, THW, THHW, THW-2	6	1	3	5	8	11	18	27	41	53	70
	4	1	1	3	6	8	13	20	30	40	52
	3	1	1	3	5	7	11	17	26	34	44
	2	1	1	2	4	6	9	14	22	29	38
	1	1	1	1	3	4	7	10	15	20	26
	1/0	0	1	1	2	3	6	8	13	17	23
	2/0	0	1	1	2	3	5	7	11	15	19
	3/0	0	1	1	1	2	4	6	9	12	16
	4/0	0	0	1	1	1	3	5	8	10	13
	250	0	0	1	1	1	3	4	6	8	11
	300	0	0	1	1	1	2	3	5	7	9
	350	0	0	0	1	1	1	3	5	6	8
	400	0	0	0	1	1	1	3	4	6	7
	500	0	0	0	1	1	1	2	3	5	6
	600	0	0	0	1	1	1	1	3	4	5
	700	0	0	0	0	1	1	1	2	3	4
	750	0	0	0	0	1	1	1	2	3	4
THHN, THWN, THWN-2	14	13	22	36	63	81	133	201	308	401	523
	12	9	16	26	46	59	97	146	225	292	381
	10	6	10	16	29	37	61	92	141	184	240
	8	3	6	9	16	21	35	53	81	106	138
	6	2	4	7	12	15	25	38	59	76	100
	4	1	2	4	7	9	15	23	36	47	61
	3	1	1	3	6	8	13	20	30	40	52
	2	1	1	3	5	7	11	17	26	33	44
	1	1	1	1	4	5	8	12	19	25	32

MAXIMUM NUMBER OF CONDUCTORS IN LIQUIDTIGHT FLEXIBLE METAL CONDUIT

Type Letters	Cond. Size AWG/kcmil	Trade Sizes In Inches ½	¾	1	1¼	1½	2	2½	3	3½	4
THHN, THWN, THWN-2	1/0	1	1	1	3	4	7	10	16	21	27
	2/0	0	1	1	2	3	6	8	13	17	23
	3/0	0	1	1	1	3	5	7	11	14	19
	4/0	0	1	1	1	2	4	6	9	12	15
	250	0	0	1	1	1	3	5	7	10	12
	300	0	0	1	1	1	3	4	6	8	11
	350	0	0	1	1	1	2	3	5	7	9
	400	0	0	0	1	1	1	3	5	6	8
	500	0	0	0	1	1	1	2	4	5	7
	600	0	0	0	1	1	1	1	3	4	6
	700	0	0	0	1	1	1	1	3	4	5
	750	0	0	0	0	1	1	1	3	3	5
FEP, FEPB, PFA, PFAH, TFE	14	12	21	35	61	79	129	195	299	389	507
	12	9	15	25	44	57	94	142	218	284	370
	10	6	11	18	32	41	68	102	156	203	266
	8	3	6	10	18	23	39	58	89	117	152
	6	2	4	7	13	17	27	41	64	83	108
	4	1	3	5	9	12	19	29	44	58	75
	3	1	2	4	7	10	16	24	37	48	63
	2	1	1	3	6	8	13	20	30	40	52
PFA, PFAH, TFE	1	1	1	2	4	5	9	14	21	28	36
PFA, PFAH, TFE, Z	1/0	1	1	1	3	4	7	11	18	23	30
	2/0	1	1	1	3	4	6	9	14	19	25
	3/0	0	1	1	2	3	5	8	12	16	20
	4/0	0	1	1	1	2	4	6	10	13	17
Z	14	20	26	42	73	95	156	235	360	469	611
	12	14	18	30	52	67	111	167	255	332	434
	10	8	11	18	32	41	68	102	156	203	266
	8	5	7	11	20	26	43	64	99	129	168
	6	4	5	8	14	18	30	45	69	90	118
	4	2	3	5	9	12	20	31	48	62	81
	3	2	2	4	7	9	15	23	35	45	59
	2	1	1	3	6	7	12	19	29	38	49
	1	1	1	2	5	6	10	15	23	30	40
XHH, XHHW, XHHW-2, ZW	14	9	15	25	44	57	93	140	215	280	365
	12	7	12	19	33	43	71	108	165	215	280
	10	5	9	14	25	32	53	80	123	160	209
	8	3	5	8	14	18	29	44	68	89	116
	6	1	3	6	10	13	22	33	50	66	86
	4	1	2	4	7	9	16	24	36	48	62
	3	1	1	3	6	8	13	20	31	40	52
	2	1	1	3	5	7	11	17	26	34	44
	1	1	1	1	4	5	8	12	19	25	33
XHH, XHHW, XHHW-2	1/0	1	1	1	3	4	7	10	16	21	28
	2/0	0	1	1	2	3	6	9	13	17	23
	3/0	0	1	1	1	3	5	7	11	14	19
	4/0	0	1	1	1	2	4	6	9	12	16
	250	0	0	1	1	1	3	5	7	10	13
	300	0	0	1	1	1	3	4	6	8	11
	350	0	0	1	1	1	2	3	5	7	10
	400	0	0	0	1	1	1	3	5	6	0
	500	0	0	0	1	1	1	2	4	5	7
	600	0	0	0	1	1	1	1	3	4	6
	700	0	0	0	1	1	1	1	3	4	5
	750	0	0	0	0	1	1	1	3	3	5

*Types RHH, RHW, and RHW-2 without outer covering.

See *Ugly's* page 137 for Trade Size / Metric Designator conversion.

Source: NFPA 70, *National Electrical Code®*, NFPA, Quincy, MA, 2013, Table C.7, as modified.

DIMENSIONS OF INSULATED CONDUCTORS AND FIXTURE WIRES

TYPE	SIZE	APPROX. AREA SQ. IN.
RFH-2	18	0.0145
FFH-2	16	0.0172
RHW-2, RHH	14	0.0293
RHW	12	0.0353
	10	0.0437
	8	0.0835
	6	0.1041
	4	0.1333
	3	0.1521
	2	0.1750
	1	0.2660
	1/0	0.3039
	2/0	0.3505
	3/0	0.4072
	4/0	0.4754
	250	0.6291
	300	0.7088
	350	0.7870
	400	0.8626
	500	1.0082
	600	1.2135
	700	1.3561
	750	1.4272
	800	1.4957
	900	1.6377
	1000	1.7719
	1250	2.3479
	1500	2.6938
	1750	3.0357
	2000	3.3719
SF-2, SFF-2	18	0.0115
	16	0.0139
	14	0.0172
SF-1, SFF-1	18	0.0065
RFH-1, XF, XFF	18	0.0080
TF, TFF, XF, XFF	16	0.0109
TW, XF, XFF, THHW, THW, THW-2	14	0.0139
TW, THHW, THW, THW-2	12	0.0181
	10	0.0243
	8	0.0437
RHH*, RHW*, RHW-2*	14	0.0209
RHH*, RHW*, RHW-2*, XF, XFF	12	0.0260

TYPE	SIZE	APPROX. AREA SQ. IN.
RHH*, RHW*, XF RHW-2*, XFF	10	0.0333
RHH*, RHW*, RHW-2*	8	0.0556
TW, THW	6	0.0726
THHW	4	0.0973
THW-2	3	0.1134
RHH*	2	0.1333
RHW*	1	0.1901
RHW-2*	1/0	0.2223
	2/0	0.2624
	3/0	0.3117
	4/0	0.3718
	250	0.4596
	300	0.5281
	350	0.5958
	400	0.6619
	500	0.7901
	600	0.9729
	700	1.1010
	750	1.1652
	800	1.2272
	900	1.3561
	1000	1.4784
	1250	1.8602
	1500	2.1695
	1750	2.4773
	2000	2.7818
TFN	18	0.0055
TFFN	16	0.0072
THHN	14	0.0097
THWN	12	0.0133
THWN-2	10	0.0211
	8	0.0366
	6	0.0507
	4	0.0824
	3	0.0973
	2	0.1158
	1	0.1562
	1/0	0.1855
	2/0	0.2223
	3/0	0.2679
	4/0	0.3237
	250	0.3970
	300	0.4608
	350	0.5242
	400	0.5863
	500	0.7073
	600	0.8676
	700	0.9887

DIMENSIONS OF INSULATED CONDUCTORS AND FIXTURE WIRES

TYPE	SIZE	APPROX. AREA SQ. IN.
THHN	750	1.0496
THWN	800	1.1085
THWN-2	900	1.2311
	1000	1.3478
PF, PGFF, PGF, PFF,	18	0.0058
PTF, PAF, PTFF, PAFF	16	0.0075
PF, PGFF, PGF, PFF,	14	0.0100
PTF, PAF, PTFF, PAFF		
TFE, FEP, PFA		
FEPB, PFAH		
TFE, FEP,	12	0.0137
PFA, FEPB,	10	0.0191
PFAH	8	0.0333
	6	0.0468
	4	0.0670
	3	0.0804
	2	0.0973
TFE, PFAH	1	0.1399
TFE,	1/0	0.1676
PFA,	2/0	0.2027
PFAH, Z	3/0	0.2463
	4/0	0.3000
ZF, ZFF	18	0.0045
	16	0.0061
Z, ZF, ZFF	14	0.0083
Z	12	0.0117
	10	0.0191
	8	0.0302
	6	0.0430
	4	0.0625
	3	0.0855
	2	0.1029
	1	0.1269
XHHW, ZW	14	0.0139
XHHW-2	12	0.0181
XHH	10	0.0243
	8	0.0437
	6	0.0590
	4	0.0814
	3	0.0962
	2	0.1146
XHHW	1	0.1534
XHHW-2	1/0	0.1825
XHH	2/0	0.2190
	3/0	0.2642
	4/0	0.3197
	250	0.3904
XHHW	300	0.4536
XHHW-2	350	0.5166
XHH	400	0.5782
	500	0.6984
	600	0.8709
	700	0.9923
	750	1.0532
	800	1.1122
	900	1.2351
	1000	1.3519
	1250	1.7180
	1500	2.0157
	1750	2.3127
	2000	2.6073
KF-2	18	0.0031
KFF-2	16	0.0044
	14	0.0064
	12	0.0093
	10	0.0139
KF-1	18	0.0026
KFF-1	16	0.0037
	14	0.0055
	12	0.0083
	10	0.0127

*Types RHH, RHW, and RHW-2 without outer covering

See *Ugly's* page 135 for conversion of square inches to mm^2.

Source: NFPA 70, *National Electrical Code*®, NFPA, Quincy, MA, 2013, Table 5, as modified.

COMPACT COPPER AND ALUMINUM BUILDING WIRE NOMINAL DIMENSIONS* AND AREAS

	Bare Conductor		Types THW and THHW		Type THHN		Type XHHW		
Size AWG or kcmil	Number of Strands	Diam. Inches	Approx. Diam. Inches	Approx. Area Sq. In.	Approx. Diam. Inches	Approx. Area Sq. In.	Approx. Diam. Inches	Approx. Area Sq. Inches	Size AWG or kcmil
8	7	0.134	0.255	0.0510	—	—	0.224	0.0394	8
6	7	0.169	0.290	0.0660	0.240	0.0452	0.260	0.0530	6
4	7	0.213	0.335	0.0881	0.305	0.0730	0.305	0.0730	4
2	7	0.268	0.390	0.1194	0.360	0.1017	0.360	0.1017	2
1	19	0.299	0.465	0.1698	0.415	0.1352	0.415	0.1352	1
1/0	19	0.336	0.500	0.1963	0.450	0.1590	0.450	0.1590	1/0
2/0	19	0.376	0.545	0.2332	0.495	0.1924	0.490	0.1885	2/0
3/0	19	0.423	0.590	0.2733	0.540	0.2290	0.540	0.2290	3/0
4/0	19	0.475	0.645	0.3267	0.595	0.2780	0.590	0.2733	4/0
250	37	0.520	0.725	0.4128	0.670	0.3525	0.660	0.3421	250
300	37	0.570	0.775	0.4717	0.720	0.4071	0.715	0.4015	300
350	37	0.616	0.820	0.5281	0.770	0.4656	0.760	0.4536	350
400	37	0.659	0.865	0.5876	0.815	0.5216	0.800	0.5026	400
500	37	0.736	0.940	0.6939	0.885	0.6151	0.880	0.6082	500
600	61	0.813	1.050	0.8659	0.985	0.7620	0.980	0.7542	600
700	61	0.877	1.110	0.9676	1.050	0.8659	1.050	0.8659	700
750	61	0.908	1.150	1.0386	1.075	0.9076	1.090	0.9331	750
900	61	0.999	1.224	1.1766	1.194	1.1196	1.169	1.0733	900
1000	61	1.060	1.285	1.2968	1.255	1.2370	1.230	1.1882	1000

*Dimensions are from industry sources.

See *Ugly's* pages 133–141 for metric conversions.

Source: NFPA 70 *National Electrical Code*®, NFPA, Quincy, MA, 2013, Table 5A, as modified.

DIMENSIONS AND PERCENT AREA OF CONDUIT AND TUBING

(For the combinations of wires permitted in Table 1, Chapter 9, *NEC®*)
(See *Ugly's* pages 133–141 for metric conversions.)

Trade Size Inches	Internal Diameter Inches	Total Area 100% Sq. in.	2 Wires 31% Sq. in.	Over 2 Wires 40% Sq. in.	1 Wire 53% Sq. in.	(NIPPLE) 60% Sq. in.
ELECTRICAL METALLIC TUBING (EMT)						
½	0.622	0.304	0.094	0.122	0.161	0.182
¾	0.824	0.533	0.165	0.213	0.283	0.320
1	1.049	0.864	0.268	0.346	0.458	0.519
1¼	1.380	1.496	0.464	0.598	0.793	0.897
1½	1.610	2.036	0.631	0.814	1.079	1.221
2	2.067	3.356	1.040	1.342	1.778	2.013
2½	2.731	5.858	1.816	2.343	3.105	3.515
3	3.356	8.846	2.742	3.538	4.688	5.307
3½	3.834	11.545	3.579	4.618	6.119	6.927
4	4.334	14.753	4.573	5.901	7.819	8.852
ELECTRICAL NONMETALLIC TUBING (ENT)						
½	0.560	0.246	0.076	0.099	0.131	0.148
¾	0.760	0.454	0.141	0.181	0.240	0.272
1	1.000	0.785	0.243	0.314	0.416	0.471
1¼	1.340	1.410	0.437	0.564	0.747	0.846
1½	1.570	1.936	0.600	0.774	1.026	1.162
2	2.020	3.205	0.993	1.282	1.699	1.923
2½	–	–	–	–	–	–
3	–	–	–	–	–	–
3½	–	–	–	–	–	–
4	–	–	–	–	–	–
FLEXIBLE METAL CONDUIT (FMC)						
⅜	0.384	0.116	0.036	0.046	0.061	0.069
½	0.635	0.317	0.098	0.127	0.168	0.190
¾	0.824	0.533	0.165	0.213	0.283	0.320
1	1.020	0.817	0.253	0.327	0.433	0.490
1¼	1.275	1.277	0.396	0.511	0.677	0.766
1½	1.538	1.858	0.576	0.743	0.985	1.115
2	2.040	3.269	1.013	1.307	1.732	1.961
2½	2.500	4.909	1.522	1.963	2.602	2.945
3	3.000	7.069	2.191	2.827	3.746	4.241
3½	3.500	9.621	2.983	3.848	5.099	5.773
4	4.000	12.566	3.896	5.027	6.660	7.540
INTERMEDIATE METAL CONDUIT (IMC)						
⅜	–	–	–	–		–
½	0.660	0.342	0.106	0.137	0.181	0.205
¾	0.864	0.586	0.182	0.235	0.311	0.352
1	1.105	0.959	0.297	0.384	0.508	0.575
1¼	1.448	1.647	0.510	0.659	0.873	0.988
1½	1.683	2.225	0.690	0.890	1.179	1.335
2	2.150	3.630	1.125	1.452	1.924	2.178
2½	2.557	5.135	1.592	2.054	2.722	3.081
3	3.176	7.922	2.456	3.169	4.199	4.753
3½	3.671	10.584	3.281	4.234	5.610	6.351
4	4.166	13.631	4.226	5.452	7.224	8.179

(continues)

DIMENSIONS AND PERCENT AREA OF CONDUIT AND TUBING

(For the combinations of wires permitted in Table 1, Chapter 9, *NEC®*)
(See *Ugly's* pages 133–141 for metric conversions.)

Trade Size Inches	Internal Diameter Inches	Total Area 100% Sq. in.	2 Wires 31% Sq. in.	Over 2 Wires 40% Sq. in.	1 Wire 53% Sq. in.	(NIPPLE) 60% Sq. in.
LIQUIDTIGHT FLEXIBLE NONMETALLIC CONDUIT (TYPE LFNC-B*)						
⅜	0.494	0.192	0.059	0.077	0.102	0.115
½	0.632	0.314	0.097	0.125	0.166	0.188
¾	0.830	0.541	0.168	0.216	0.287	0.325
1	1.054	0.873	0.270	0.349	0.462	0.524
1¼	1.395	1.528	0.474	0.611	0.810	0.917
1½	1.588	1.981	0.614	0.792	1.050	1.188
2	2.033	3.246	1.006	1.298	1.720	1.948
*Corresponds to Section 356.2(2).						
LIQUIDTIGHT FLEXIBLE NONMETALLIC CONDUIT (TYPE LFNC-A*)						
⅜	0.495	0.192	0.060	0.077	0.102	0.115
½	0.630	0.312	0.097	0.125	0.165	0.187
¾	0.825	0.535	0.166	0.214	0.283	0.321
1	1.043	0.854	0.265	0.342	0.453	0.513
1¼	1.383	1.502	0.466	0.601	0.796	0.901
1½	1.603	2.018	0.626	0.807	1.070	1.211
2	2.063	3.343	1.036	1.337	1.772	2.006
*Corresponds to Section 356.2(1).						
LIQUIDTIGHT FLEXIBLE METAL CONDUIT (LFMC)						
⅜	0.494	0.192	0.059	0.077	0.102	0.115
½	0.632	0.314	0.097	0.125	0.166	0.188
¾	0.830	0.541	0.168	0.216	0.287	0.325
1	1.054	0.873	0.270	0.349	0.462	0.524
1¼	1.395	1.528	0.474	0.611	0.810	0.917
1½	1.588	1.981	0.614	0.792	1.050	1.188
2	2.033	3.246	1.006	1.298	1.720	1.948
2½	2.493	4.881	1.513	1.953	2.587	2.929
3	3.085	7.475	2.317	2.990	3.962	4.485
3½	3.520	9.731	3.017	3.893	5.158	5.839
4	4.020	12.692	3.935	5.077	6.727	7.615
RIGID METAL CONDUIT (RMC)						
⅜	–	–	–	–	–	–
½	0.632	0.314	0.097	0.125	0.166	0.188
¾	0.836	0.549	0.170	0.220	0.291	0.329
1	1.063	0.887	0.275	0.355	0.470	0.532
1¼	1.394	1.526	0.473	0.610	0.809	0.916
1½	1.624	2.071	0.642	0.829	1.098	1.243
2	2.083	3.408	1.056	1.363	1.806	2.045
2½	2.489	4.866	1.508	1.946	2.579	2.919
3	3.090	7.499	2.325	3.000	3.974	4.499
3½	3.570	10.010	3.103	4.004	5.305	6.006
4	4.050	12.882	3.994	5.153	6.828	7.729
5	5.073	20.212	6.266	8.085	10.713	12.127
6	6.093	29.158	9.039	11.663	15.454	17.495

DIMENSIONS AND PERCENT AREA OF CONDUIT AND TUBING

(For the combinations of wires permitted in Table 1, Chapter 9, *NEC*®)
(See *Ugly's* pages 133–141 for metric conversions.)

Trade Size Inches	Internal Diameter Inches	Total Area 100% Sq. in.	2 Wires 31% Sq. in.	Over 2 Wires 40% Sq. in.	1 Wire 53% Sq. in.	(NIPPLE) 60% Sq. in.
RIGID PVC CONDUIT (PVC), SCHEDULE 80						
½	0.526	0.217	0.067	0.087	0.115	0.130
¾	0.722	0.409	0.127	0.164	0.217	0.246
1	0.936	0.688	0.213	0.275	0.365	0.413
1¼	1.255	1.237	0.383	0.495	0.656	0.742
1½	1.476	1.711	0.530	0.684	0.907	1.027
2	1.913	2.874	0.891	1.150	1.523	1.725
2½	2.290	4.119	1.277	1.647	2.183	2.471
3	2.864	6.442	1.997	2.577	3.414	3.865
3½	3.326	8.688	2.693	3.475	4.605	5.213
4	3.786	11.258	3.490	4.503	5.967	6.755
5	4.768	17.855	5.535	7.142	9.463	10.713
6	5.709	25.598	7.935	10.239	13.567	15.359
RIGID PVC CONDUIT (PVC), SCHEDULE 40 & HDPE CONDUIT (HDPE)						
½	0.602	0.285	0.088	0.114	0.151	0.171
¾	0.804	0.508	0.157	0.203	0.269	0.305
1	1.029	0.832	0.258	0.333	0.441	0.499
1¼	1.360	1.453	0.450	0.581	0.770	0.872
1½	1.590	1.986	0.616	0.794	1.052	1.191
2	2.047	3.291	1.020	1.316	1.744	1.975
2½	2.445	4.695	1.455	1.878	2.488	2.817
3	3.042	7.268	2.253	2.907	3.852	4.361
3½	3.521	9.737	3.018	3.895	5.161	5.842
4	3.998	12.554	3.892	5.022	6.654	7.532
5	5.016	19.761	6.126	7.904	10.473	11.856
6	6.031	28.567	8.856	11.427	15.141	17.140
TYPE A, RIGID PVC CONDUIT (PVC)						
½	0.700	0.385	0.119	0.154	0.204	0.231
¾	0.910	0.650	0.202	0.260	0.345	0.390
1	1.175	1.084	0.336	0.434	0.575	0.651
1¼	1.500	1.767	0.548	0.707	0.937	1.060
1½	1.720	2.324	0.720	0.929	1.231	1.394
2	2.155	3.647	1.131	1.459	1.933	2.188
2½	2.635	5.453	1.690	2.181	2.890	3.272
3	3.230	8.194	2.540	3.278	4.343	4.916
3½	3.690	10.694	3.315	4.278	5.668	6.416
4	4.180	13.723	4.254	5.409	7.273	8.234
TYPE EB, PVC CONDUIT (PVC)						
2	2.221	3.874	1.201	1.550	2.053	2.325
2½	–	–	–	–	–	–
3	3.330	8.709	2.700	3.484	4.616	5.226
3½	3.804	11.365	3.523	4.546	6.023	6.819
4	4.289	14.448	4.479	5.779	7.657	8.669
5	5.316	22.195	6.881	8.878	11.763	13.317
6	6.336	31.530	9.774	12.612	16.711	18.918

Source: NFPA 70, *National Electrical Code*®, NFPA, Quincy, MA, 2013, Table 4, as modified.

THREAD DIMENSIONS AND TAP DRILL SIZES

COARSE THREAD SERIES				FINE THREAD SERIES			
NOMINAL SIZE	THREADS PER IN.	TAP DRILL	CLEARANCE DRILL	NOMINAL SIZE	THREADS PER IN.	TAP DRILL	CLEARANCE DRILL
5/64"	48	47	36	0	80	3/64"	51
1/8	40	38	29	1	72	53	47
6	32	36	25	2	64	50	42
8	32	29	16	3	56	45	36
10	24	25	13/64"	4	48	42	31
12	24	16	7/32"	1/8"	44	37	29
1/4"	20	7	17/64"	6	40	33	25
5/16"	18	F	21/64"	8	36	29	16
3/8"	16	5/16"	25/64"	10	32	21	13/64"
7/16"	14	U	29/64"	12	28	14	7/32"
1/2"	13	27/64"	33/64"	1/4"	28	3	17/64"
9/16"	12	31/64"	37/64"	5/16"	24	1	21/64"
5/8"	11	17/32"	41/64"	3/8"	24	Q	25/64"
3/4"	10	21/32"	49/64"	7/16"	20	25/64"	29/64"
7/8"	9	49/64"	57/64"	1/2"	20	29/64"	33/64"
1"	8	7/8"	1-1/64"	9/16"	18	33/64"	37/64"
1-1/4"	7	1-11/64"	1-17/64"	5/8"	18	37/64"	41/64"
1-3/8"	6	1-19/64"	1-25/64"	3/4"	16	11/16"	49/64"
1-1/2"	6	1-27/64"	1-33/64"	7/8"	14	13/16"	57/64"
2"	4-1/2"	1-25/32"	2-1/32"	1	14	15/16"	1-1/64"

HOLE SAW CHART*

TRADE SIZE	RIGID CONDUIT	E.M.T. CONDUIT	GREEN-FIELD	L.T. FLEX.	TRADE SIZE	RIGID CONDUIT	E.M.T. CONDUIT	GREEN-FIELD
1/2"	7/8"	3/4"	1"	1-1/8"	2-1/2"	3"	2-7/8"	2-7/8"
3/4	1-1/8"	1"	1-1/8"	1-1/4"	3"	3-5/8"	3-1/2"	3-5/8"
1"	1-3/8"	1-1/4"	1-1/2"	1-1/2"	3-1/2"	4-1/8"	4"	4-1/8"
1-1/4"	1-3/4"	1-5/8"	1-3/4"	1-7/8"	4"	4-5/8"	4-1/2"	4-5/8"
1-1/2"	2"	1-7/8"	2"	2-1/8"	5"	5-3/4"		
2"	2-1/2"	2-1/8"	2-1/2"	2-3/4"	6"	6-3/4"		

NOTE: For oil-type push button station, use size 1-7/32" knock-out punch.
* For connectors (male connectors and adapters), use Rigid Table.

METAL BOXES

BOX DIMENSION, INCHES TRADE SIZE, OR TYPE	MIN. CU. IN. CAPACITY	MAXIMUM NUMBER OF CONDUCTORS						
		NO. 18	NO. 16	NO. 14	NO. 12	NO. 10	NO. 8	NO. 6
4 x 1-1/4 ROUND OR OCTAGONAL	12.5	8	7	6	5	5	4	2
4 x 1-1/2 ROUND OR OCTAGONAL	15.5	10	8	7	6	6	5	3
4 x 2-1/8 ROUND OR OCTAGONAL	21.5	14	12	10	9	8	7	4
4 x 1-1/4 SQUARE	18.0	12	10	9	8	7	6	3
4 x 1-1/2 SQUARE	21.0	14	12	10	9	8	7	4
4 x 2-1/8 SQUARE	30.3	20	17	15	13	12	10	6
4-11/16 x 1-1/4 SQUARE	25.5	17	14	12	11	10	8	5
4-11/16 x 1-1/2 SQUARE	29.5	19	16	14	13	11	9	5
4-11/16 x 2-1/8 SQUARE	42.0	28	24	21	18	16	14	8
3 x 2 x 1-1/2 DEVICE	7.5	5	4	3	3	3	2	1
3 x 2 x 2 DEVICE	10.0	6	5	5	4	4	3	2
3 x 2 x 2-1/4 DEVICE	10.5	7	6	5	4	4	3	2
3 x 2 x 2-1/2 DEVICE	12.5	8	7	6	5	5	4	2
3 x 2 x 2-3/4 DEVICE	14.0	9	8	7	6	5	4	2
3 x 2 x 3-1/2 DEVICE	18.0	12	10	9	8	7	6	3
4 x 2-1/8 x 1-1/2 DEVICE	10.3	6	5	5	4	4	3	2
4 x 2-1/8 x 1-7/8 DEVICE	13.0	8	7	6	5	5	4	2
4 x 2-1/8 x 2-1/8 DEVICE	14.5	9	8	7	6	5	4	2
3-3/4 x 2 x 2-1/2 MASONRY BOX/GANG	14.0	9	8	7	6	5	4	2
3-3/4 x 2 x 3-1/2 MASONRY BOX/GANG	21.0	14	12	10	9	8	7	4
FS-MINIMUM INTERNAL DEPTH 1-3/4 SINGLE COVER/GANG	13.5	9	7	6	6	5	4	2
FD-MINIMUM INTERNAL DEPTH 2-3/8 SINGLE COVER/GANG	18.0	12	10	9	8	7	6	3
FS-MINIMUM INTERNAL DEPTH 1-3/4 MULTIPLE COVER/GANG	18.0	12	10	9	8	7	6	3
FD-MINIMUM INTERNAL DEPTH 2-3/8 MULTIPLE COVER/GANG	24.0	16	13	12	10	9	8	4

Source: NFPA 70, *National Electrical Code*®, NFPA, Quincy, MA, 2013, Table 314.16(A), as modified.

MINIMUM COVER REQUIREMENTS, 0–1000 VOLTS, NOMINAL

Cover is defined as the distance between the top surface of direct burial cable, conduit, or other raceways and the finished surface.

WIRING METHOD	MINIMUM BURIAL (INCHES)
DIRECT BURIAL CABLES	24
RIGID METAL CONDUIT	6*
INTERMEDIATE METAL CONDUIT	6*
RIGID NONMETALLIC CONDUIT (APPROVED FOR DIRECT BURIAL WITHOUT CONCRETE ENCASEMENT)	18*

*For most locations, for complete details, refer to *NEC*® Table 300.5 for exceptions such as highways, dwellings, airports, driveways, parking lots, etc.
Source: Data from NFPA 70, *National Electrical Code*®, NFPA, Quincy, MA, 2013, Table 300.5.

VOLUME REQUIRED PER CONDUCTOR

SIZE OF CONDUCTOR	FREE SPACE WITHIN BOX FOR EACH CONDUCTOR
No. 18	1.5 CUBIC INCHES
No. 16	1.75 CUBIC INCHES
No. 14	2 CUBIC INCHES
No. 12	2.25 CUBIC INCHES
No. 10	2.5 CUBIC INCHES
No. 8	3 CUBIC INCHES
No. 6	5 CUBIC INCHES

For complete details see *NEC* 314.16(B).
Source: NFPA 70, *National Electrical Code*®, NFPA, Quincy, MA, 2013, Table 314.16(B), as modified.

SPACINGS FOR CONDUCTOR SUPPORTS

Conductor Size	Support of Conductors in Vertical Raceways	CONDUCTORS Aluminum or Copper-Clad Aluminum	Copper
18 AWG through 8 AWG	Not greater than	100 feet	100 feet
6 AWG through 1/0 AWG	Not greater than	200 feet	100 feet
2/0 AWG through 4/0 AWG	Not greater than	180 feet	80 feet
Over 4/0 AWG through 350 kcmil	Not greater than	135 feet	60 feet
Over 350 kcmil through 500 kcmil	Not greater than	120 feet	50 feet
Over 500 kcmil through 750 kcmil	Not greater than	95 feet	40 feet
Over 750 kcmil	Not greater than	85 feet	35 feet

For SI units: 1 foot = 0.3048 meter.
Source: NFPA 70, *National Electrical Code*®, NFPA, Quincy, MA, 2013, Table 300.19(A), as modified.

MINIMUM DEPTH OF CLEAR WORKING SPACE AT ELECTRICAL EQUIPMENT

NOMINAL VOLTAGE TO GROUND	CONDITIONS 1	2	3
	Minimum clear distance (ft)		
0–150 V	3	3	3
151–600 V	3	3½	4
601–2500 V	3	4	5
2501–9000 V	4	5	6
9001–25,000 V	5	6	9
25,001 V–75 kV	6	8	10
Above 75 kV	8	10	12

NOTES:

1. For SI units, 1 ft = 0.3048 m.
2. Where the conditions are as follows:
   Condition 1 – Exposed live parts on one side of the working space and no live or grounded parts on the other side of the working space, or exposed live parts on both sides of the working space that are effectively guarded by insulating materials.
   Condition 2 – Exposed live parts on one side of the working space and grounded parts on the other side of the working space. Concrete, brick, or tile walls shall be considered as grounded.
   Condition 3 – Exposed live parts on both sides of the working space.

See *Ugly's* pages 133–141 for metric conversions. For electrical rooms where the equipment is rated 800 amps or more, *NEC* Article 110.26(C)(3) requires personnel doors to open in the egress direction and be equipped with listed panic hardware. Entrance to rooms shall meet the requirements of Article 110.28(C) and shall be clearly marked with warning signs forbidding unqualified persons to enter.

Source: NFPA 70, *National Electrical Code®*, NFPA, Quincy, MA, 2013, Tables 110.26(A)(1) and 110.34(A), as modified.

MINIMUM CLEARANCE OF LIVE PARTS

NOMINAL VOLTAGE RATING KV	IMPULSE WITHSTAND B.I.L. KV		MINIMUM CLEARANCE OF LIVE PARTS, INCHES*			
			PHASE-TO-PHASE		PHASE-TO-GROUND	
	INDOORS	OUTDOORS	INDOORS	OUTDOORS	INDOORS	OUTDOORS
2.4–4.16	60	95	4.5	7	3.0	6
7.2	75	95	5.5	7	4.0	6
13.8	95	110	7.5	12	5.0	7
14.4	110	110	9.0	12	6.5	7
23	125	150	10.5	15	7.5	10
34.5	150	150	12.5	15	9.5	10
	200	200	18.0	18	13.0	13
46		200		18		13
		250		21		17
69		250		21		17
		350		31		25
115		550		53		42
138		550		53		42
		650		63		50
161		650		63		50
		750		72		58
230		750		72		58
		900		89		71
		1050		105		83

For SI units: 1 inch = 25.4 millimeters

* The values given are the minimum clearance for rigid parts and bare conductors under favorable service conditions. They shall be increased for conductor movement or under unfavorable service conditions, or wherever space limitations permit. The selection of the associated impulse withstand voltage for a particular system voltage is determined by the characteristics of the surge protective equipment.

See *Ugly's* pages 133–141 for metric conversions.

Source: NFPA 70, *National Electrical Code*®, NFPA, Quincy, MA, 2013, Table 490.24, as modified.

MINIMUM SIZE EQUIPMENT GROUNDING CONDUCTORS FOR GROUNDING RACEWAY AND EQUIPMENT

RATING OR SETTING OF AUTOMATIC OVERCURRENT DEVICE IN CIRCUIT AHEAD OF EQUIPMENT, CONDUIT, ETC., NOT EXCEEDING (AMPERES)	SIZE	
	COPPER	ALUMINUM OR COPPER-CLAD ALUMINUM*
15	14	12
20	12	10
30	10	8
40	10	8
60	10	8
100	8	6
200	6	4
300	4	2
400	3	1
500	2	1/0
600	1	2/0
800	1/0	3/0
1000	2/0	4/0
1200	3/0	250 kcmil
1600	4/0	350 kcmil
2000	250 kcmil	400 kcmil
2500	350 kcmil	600 kcmil
3000	400 kcmil	600 kcmil
4000	500 kcmil	750 kcmil
5000	700 kcmil	1200 kcmil
6000	800 kcmil	1200 kcmil

NOTE: Where necessary to comply with *NEC* Section 250.4(A)(5) or 250.4(B)(4), the equipment grounding conductor shall be sized larger than given in this table.

* See installation restrictions in *NEC* 250.120.

Source: NFPA 70, *National Electrical Code*®, NFPA, Quincy, MA, 2013, Table 250.122, as modified.

GROUNDING ELECTRODE CONDUCTOR FOR ALTERNATING-CURRENT SYSTEMS

SIZE OF LARGEST UNGROUNDED SERVICE-ENTRANCE CONDUCTOR OR EQUIVALENT AREA FOR PARALLEL CONDUCTORS*		SIZE OF GROUNDING ELECTRODE CONDUCTOR	
COPPER	ALUMINUM OR COPPER-CLAD ALUMINUM	COPPER	ALUMINUM OR COPPER-CLAD ALUMINUM**
2 OR SMALLER	1/0 OR SMALLER	8	6
1 OR 1/0	2/0 OR 3/0	6	4
2/0 OR 3/0	4/0 OR 250 kcmil	4	2
OVER 3/0 THRU 350 kcmil	OVER 250 THRU THRU 500 kcmil	2	1/0
OVER 350 kcmil THRU 600 kcmil	OVER 500 kcmil THRU 900 kcmil	1/0	3/0
OVER 600 kcmil THRU 1100 kcmil	OVER 900 kcmil THRU 1750 kcmil	2/0	4/0
OVER 1100 kcmil	OVER 1750 kcmil	3/0	250 kcmil

NOTES:

1. Where multiple sets of service-entrance conductors are used as permitted in *NEC* Section 230.40, Exception No. 2, the equivalent size of the largest service-entrance conductor shall be determined by the largest sum of the areas of the corresponding conductors of each set.
2. Where there are no service-entrance conductors, the grounding electrode conductor size shall be determined by the equivalent size of the largest service-entrance conductor required for the load to be served.

* This table also applies to the derived conductors of separately derived ac systems.

** See installation restrictions in *NEC* Section 250.64(A).

INFORMATIONAL NOTE: See *NEC* Section 250.24(C) for size of ac system conductor brought to service equipment.

Source: NFPA 70, *National Electrical Code*®, NFPA, Quincy, MA, 2013, Table 250.66, as modified.

GENERAL LIGHTING LOADS BY OCCUPANCY

TYPE OF OCCUPANCY	VOLT-AMPERES/ SQUARE FOOT
Armories & auditoriums	1
Banks	3½[b]
Barber shops & beauty parlors	3
Churches	1
Clubs	2
Court rooms	2
Dwelling units[a]	3
Garages - commercial (storage)	½
Hospitals	2
Hotels & motels, including apartment houses without provision for cooking by tenants[a]	2
Industrial commercial (loft) buildings	2
Lodge rooms	1½
Office buildings	3½[b]
Restaurants	2
Schools	3
Stores	3
Warehouses (storage)	¼
In any of the preceding occupancies except one-family dwellings & individual dwelling units of two-family & multi-family dwellings:	
Assembly halls & auditoriums	1
Halls, corridors, closets, stairways	½
Storage spaces	¼

[a] See *NEC* 220.14(J).
[b] See *NEC* 220.14(K).

Source: NFPA 70, *National Electrical Code*®, NFPA, Quincy, MA, 2013, Table 220.12, as modified.

LIGHTING LOAD DEMAND FACTORS

Type of Occupancy	Portion of Lighting Load to Which Demand Factor Applies (Volt-Amperes)	Demand Factor (Percent)
Dwelling units	First 3000 or less at	100
	From 3001 to 120,000 at	35
	Remainder over 120,000 at	25
Hospitals*	First 50,000 or less at	40
	Remainder over 50,000 at	20
Hotels and motels, including apartment houses without provision for cooking by tenants*	First 20,000 or less at	50
	From 20,001 to 100,000 at	40
	Remainder over 100,000 at	30
Warehouses (storage)	First 12,500 or less at	100
	Remainder over 12,500 at	50
All others	Total volt-amperes	100

* The demand factors of this table shall not apply to the calculated load of feeders or services supplying areas in hospitals, hotels, and motels where the entire lighting is likely to be used at one time, as in operating rooms, ballrooms, or dining rooms.

Source: NFPA 70, *National Electrical Code*®, NFPA, Quincy, MA, 2013, Table 220.42, as modified.

DEMAND FACTORS FOR NONDWELLING RECEPTACLE LOADS

Portion of Receptacle Load to Which Demand Factor Applies (Volt-Amperes)	Demand Factor (Percent)
First 10 kVA or less at	100
Remainder over 10 kVA at	50

Source: NFPA 70, *National Electrical Code®*, NFPA, Quincy, MA, 2013, Table 220.44, as modified.

DEMAND FACTORS FOR HOUSEHOLD ELECTRIC CLOTHES DRYERS

Number of Dryers	Demand Factor (Percent)
1–4	100%
5	85%
6	75%
7	65%
8	60%
9	55%
10	50%
11	47%
12–23	47% minus 1% for each dryer exceeding 11
24–42	35% minus 0.5% for each dryer exceeding 23
43 and over	25%

Source: NFPA 70, *National Electrical Code®*, NFPA, Quincy, MA, 2013, Table 220.54, as modified.

DEMAND FACTORS FOR KITCHEN EQUIPMENT - OTHER THAN DWELLING UNIT(S)

Number of Units of Equipment	Demand Factor (Percent)
1	100
2	100
3	90
4	80
5	70
6 and over	65

NOTE: In no case shall the feeder or service calculated load be less than the sum of the largest two kitchen equipment loads.

Source: NFPA 70, *National Electrical Code®*, NFPA, Quincy, MA, 2013, Table 220.56, as modified.

DEMAND LOADS FOR HOUSEHOLD ELECTRIC RANGES, WALL-MOUNTED OVENS, COUNTER-MOUNTED COOKING UNITS, AND OTHER HOUSEHOLD COOKING APPLIANCES OVER 1¾ KW RATING

(Column C to be used in all cases except as otherwise permitted in Note 3)

Number of Appliances	Demand Factor (Percent) (See Notes) Column A (less than 3½ kW Rating)	Column B (3½ kW through 8¾ kW Rating)	Column C Maximum Demand (kW) (See Notes) (not over 12 kW Rating)
1	80	80	8
2	75	65	11
3	70	55	14
4	66	50	17
5	62	45	20
6	59	43	21
7	56	40	22
8	53	36	23
9	51	35	24
10	49	34	25
11	47	32	26
12	45	32	27
13	43	32	28
14	41	32	29
15	40	32	30
16	39	28	31
17	38	28	32
18	37	28	33
19	36	28	34
20	35	28	35
21	34	26	36
22	33	26	37
23	32	26	30
24	31	26	39
25	30	26	40
26–30	30	24	15 kW + 1 kW for each range
31–40	30	22	
41–50	30	20	25 kW + 3/4 kW for each range
51–60	30	18	
61 and over	30	16	

(continues)

DEMAND LOADS FOR HOUSEHOLD ELECTRIC RANGES, WALL-MOUNTED OVENS, COUNTER-MOUNTED COOKING UNITS, AND OTHER HOUSEHOLD COOKING APPLIANCES OVER 1¾ KW RATING

NOTES:

1. Over 12 kW through 27 kW ranges all of same rating. For ranges individually rated more than 12 kW but not more than 27 kW, the maximum demand in Column C shall be increased 5% for each additional kilowatt of rating or major fraction thereof by which the rating of individual ranges exceeds 12 kW.
2. Over 8¾ kW through 27 kW ranges of unequal ratings. For ranges individually rated more than 8¾ kW and of different ratings, but none exceeding 27 kW, an average value of rating shall be computed by adding together the ratings of all ranges to obtain the total connected load (using 12 kW for any range rated less than 12 kW) and dividing the total number of ranges. Then the maximum demand in Column C shall be increased 5 percent for each kilowatt or major fraction thereof by which this average value exceeds 12 kW.
3. Over 1¾ kW through 8¾ kW. In lieu of the method provided in Column C, it shall be permissible to add the nameplate ratings of all household cooking appliances rated more than 1¾ kW but not more than 8¾ kW and multiply the sum by the demand factors specified in Column A or B for the given number of appliances. Where the rating of cooking appliances falls under both Column A and Column B, the demand factors for each column shall be applied to the appliances for that column and the results added together.
4. Branch-circuit load. It shall be permissible to compute the branch-circuit load for one range in accordance with Table 220.19. The branch-circuit load for one wall-mounted oven or one counter-mounted cooking unit shall be the nameplate rating of the appliance. The branch-circuit load for a counter-mounted cooking unit and not more than two wall-mounted ovens, all supplied from a single branch circuit and located in the same room, shall be computed by adding the nameplate rating of the individual appliances and treating this total as equivalent to one range.
5. This table also applies to household cooking appliances rated over 1¾ kW and used in instructional programs.

Source: NFPA 70, *National Electrical Code®*, NFPA, Quincy, MA, 2013, Table 220.55, as modified.

CALCULATING COST OF OPERATING AN ELECTRICAL APPLIANCE

What is the monthly cost of operating a 240-volt, 5-kilowatt (kW) central electric heater that operates 12 hours per day when the cost is 15 cents per kilowatt-hour (kWhr)?

Cost = Watts × Hours Used × Rate per kWhr / 1000

5 kW = 5000 Watts

Hours = 12 Hours × 30 Days = 360 Hours per Month

= 5000 × 360 × 0.15 / 1,000

= 270,000 / 1,000 = **$270 Monthly Cost**

The above example is for a resistive load. Air-conditioning loads are primarily inductive loads. However, if ampere and voltage values are known, this method will give an approximate cost. Kilowatt-hour rates vary for different power companies, and for residential use, graduated-rate scales are usually used (the more power used, the lower the rate). Commercial and industrial rates are generally based on kilowatt usage, maximum demand and power factor. Other costs are often added such as fuel cost adjustments.

CHANGING INCANDESCENT LAMP TO ENERGY-SAVING LAMP

A 100-watt incandescent lamp is to be replaced with a 26-watt, energy-saving lamp that has the same light output (lumens). If the cost per kilowatt-hour (kWhr) is 15 cents, how many hours would the new lamp need to operate to pay for itself?

Lamp cost is 3 dollars. Energy saved is 74 watts.

Hours = Lamp Cost × 1000 / Watts Saved × kWhr

3 × 1000 / 74 × 0.15 = 3000 / 11.1 = 270.27 hours

The energy-saving lamp will pay for itself with 270.27 hours of operation.

The comparative operating cost of these two lamps based on 270.27 hours is found by:

Cost = Watts × Hours Used × Rate per kWhr / 1000

100-watt incandescent lamp = $4.05 for 270.27 hours of operation
26-watt energy saving lamp = $1.05 for 270.27 hours of operation

PARTIAL 2014 *NEC* CODE CHANGE SUMMARY

Electrical Service Areas [*NEC* 210.64]

A minimum of one 125-volt, 15-amp receptacle shall be installed within 50 feet of the electrical service equipment, except for in one- and two-family dwellings.

Noninstantaneous Trip [*NEC* 240.87]

Noninstantaneous trip breakers shall have documentation available regarding the design, installation, operation, inspection, and location of the circuit breaker(s). One of the following methods or an approved equivalent means shall be used to reduce clearing time: zone-selective interlocking, differential relaying, energy-reducing maintenance switching with local status indicator, or energy-reducing active arc-flash mitigation system.

Direct-Current Ground-Fault Detection [*NEC* 250.167]

Ground-fault detection systems shall be required for underground systems and permitted for ground systems. Grounding type shall be marked at the dc source or first disconnecting means.

Receptacle Mounting [*NEC* 406.5(F)]

Receptacles installed in a face-up position in seating areas shall be banned unless they are part of an assembly listed for the application.

Top-Terminal Batteries [*NEC* 480.0(D)]

When top-terminal batteries are installed in tiered racks, the working space shall be in accordance with the battery manufacturer's instructions.

Battery Rooms [*NEC* 480.9(E)]

Personnel doors for battery rooms are required to open in the egress direction and be equipped with listed panic hardware.

PARTIAL 2014 *NEC* CODE CHANGE SUMMARY

Backfeed on Equipment Over 1000 Volts [*NEC* 490.25]

When the possibility of backfeed exists, installations require a permanent sign in accordance with Article 110.21(B), and a permanent single-line diagram of the switching arrangement shall be installed within sight of each point of connection.

Over 1000 Volts, Substations [*NEC* 490.48]

Documentation of the engineered design by a qualified licensed professional engineer primarily in the design of substations must be available upon request.

Operating Room Receptacles [*NEC* 517.19(C)]

A minimum of 36 receptacles are required for operating rooms, with at least 12 on either normal or critical branch circuits.

Electric Vehicle Charging System [*NEC* 625]

Article 625 has been reworked to clarify the requirements for electric vehicle charging systems.

Module Data Centers [*NEC* 646]

New Article 646 has been added to address the requirements for modular data centers.

Turbine Shutdown [*NEC* 694.23]

Article 694.23 clarifies the manual shutdown procedure requirements for turbine shutdown.

Optional Standby Systems Power Inlet [*NEC* 702.7(C)]

For power inlets for a temporary connection for a portable generator, a warning sign must be installed stating if the neutral is bonded or floating.

PARTIAL 2014 *NEC* CODE CHANGE SUMMARY

Interconnected Electric Power Production Sources [*NEC* 705.31]

Overcurrent protection for electric power production source conductors shall be located within 10 feet of the connected service.

Fire-Resistive Cable Systems [*NEC* 728]

New Article 728 has been added to address the requirements for fire-resistive cable systems.

Energy Management Systems [*NEC* 750]

New Article 750 has been added to address the requirements for energy management systems.

Annex J

New Informative Annex J ADA Standards for Accessible Design has been added as part of the 2010 ADA Standards for Accessible Design.

FIELD TERMS VERSUS *NEC* TERMS

Field term	NEC term
• BX	Armored cable (*NEC* 320)
• Romex	Non-metallic sheathed cable (*NEC* 334)
• Green field	Flexible metal conduit (*NEC* 348)
• Thin wall	Electrical metallic tubing (*NEC* 358)
• Smurf tube	Electrical non-metallic tubing (*NEC* 362)
• 1900 box	4-inch square box (*NEC* 314)
• 333 box	Device box (*NEC* 314)
• EYS	Explosion proof seal off (*NEC* 500)
• Neutral**	Grounded conductor (*NEC* 200)**
• Ground wire	Equipment grounding conductor (*NEC* 250.118)
• Ground wire	Grounding electrode conductor (*NEC* 250.66)
• Hot, live	Ungrounded conductor (*NEC* 100)

** Some systems do not have a neutral and the grounded conductor may be a phase conductor. (See *NEC* Article 100 neutral definition.)

ELECTRICAL SYMBOLS

WALL	CEILING	
		OUTLET
D	D	DROP CORD
F	F	FAN OUTLET
J	J	JUNCTION BOX
L	L	LAMP HOLDER
L_{PS}	L_{PS}	LAMP HOLDER WITH PULL SWITCH
S	S	PULL SWITCH
V	V	VAPOR DISCHARGE SWITCH
X	X	EXIT OUTLET
C	C	CLOCK OUTLET
B	B	BLANKED OUTLET
		DUPLEX CONVENIENCE OUTLET
1, 3		SINGLE, TRIPLEX, ETC.
		RANGE OUTLET
S		SWITCH AND CONVENIENCE OUTLET
		SPECIAL PURPOSE OUTLET
		FLOOR OUTLET

SWITCH OUTLETS	
S	SINGLE POLE SWITCH
S_2	DOUBLE POLE SWITCH
S_3	THREE-WAY SWITCH
S_4	FOUR-WAY SWITCH
S_D	AUTOMATIC DOOR SWITCH
S_E	ELECTROLIER SWITCH
S_P	SWITCH AND PILOT LAMP
S_K	KEY-OPERATED SWITCH
S_{CB}	CIRCUIT BREAKER
S_{WCB}	WEATHER PROOF CIRCUIT BREAKER
S_{MC}	MOMENTARY CONTACT SWITCH
S_{RC}	REMOTE CONTROL SWITCH
S_{WP}	WEATHER-PROOF SWITCH
S_F	FUSED SWITCH
S_{WPF}	WEATHER-PROOF FUSED SWITCH
	LIGHTING SWITCH
	POWER PANEL

ELECTRICAL SYMBOLS

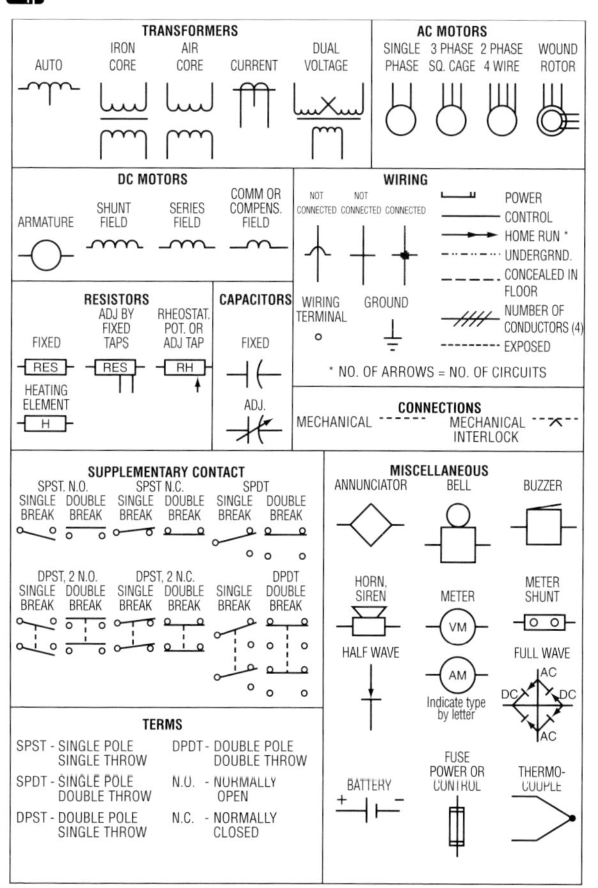

ELECTRICAL SYMBOLS

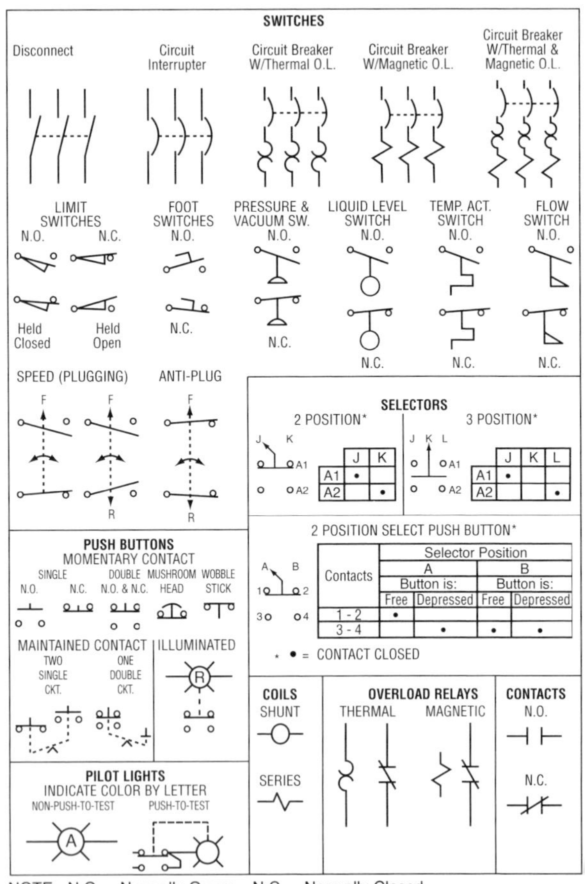

NOTE: N.O. = Normally Open; N.C. = Normally Closed

WIRING DIAGRAMS FOR NEMA CONFIGURATIONS

2 Pole, 2 Wire Nongrounding 125V

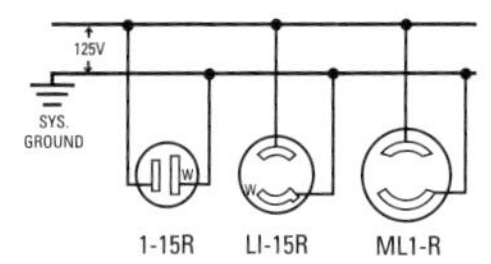

2 Pole, 2 Wire Nongrounding 250V

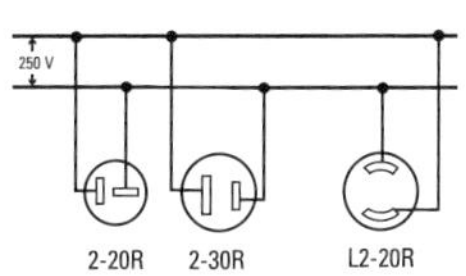

2 Pole, 3 Wire Grounding 125V

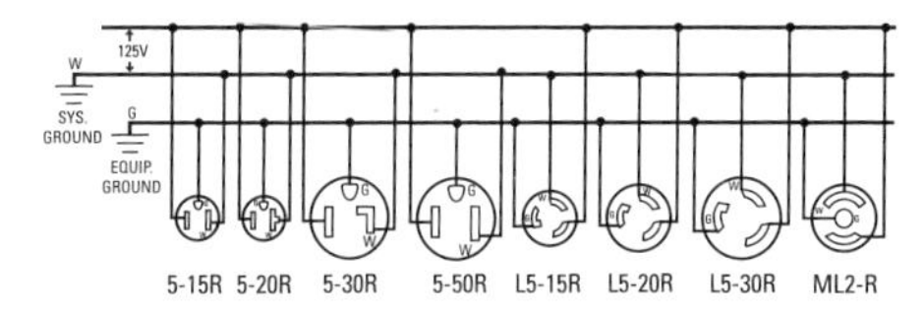

2 Pole, 3 Wire Grounding 250V

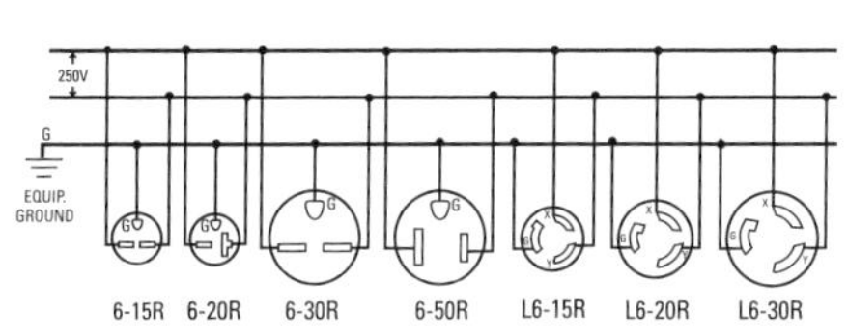

2 Pole, 3 Wire Grounding 277V AC

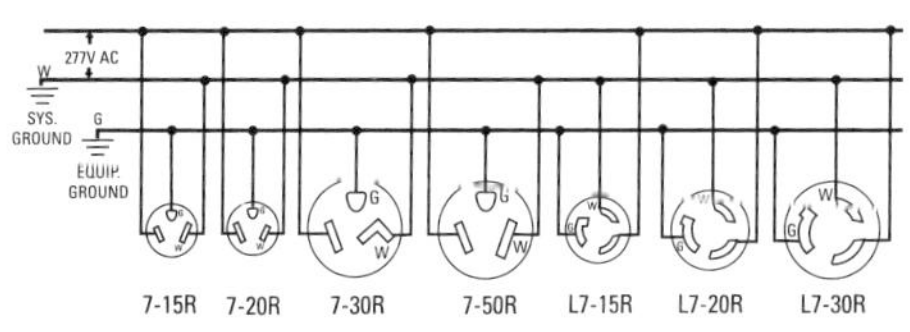

Courtesy of
COOPER Wiring Devices
The New Power in Wiring Devices

WIRING DIAGRAMS FOR NEMA CONFIGURATIONS

2 Pole, 3 Wire
Grounding
480V AC

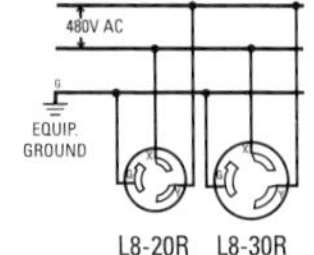

3 Pole, 3 Wire
Nongrounding
125/250V

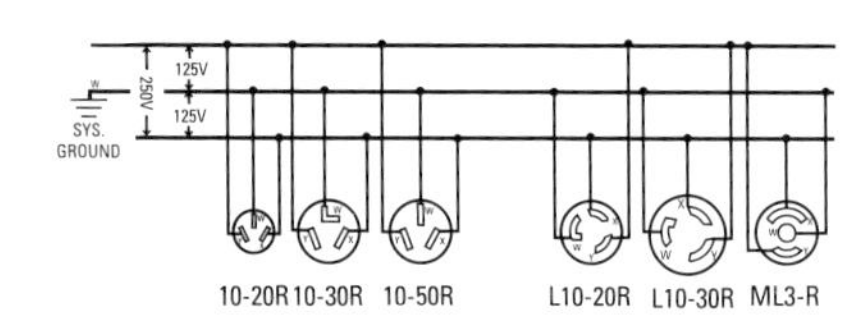

3 Pole, 3 Wire
Nongrounding
3ø 250V

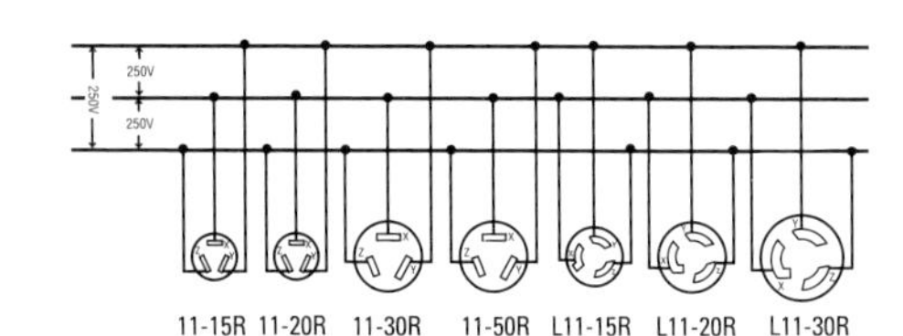

3 Pole, 4 Wire
Grounding
125/250V

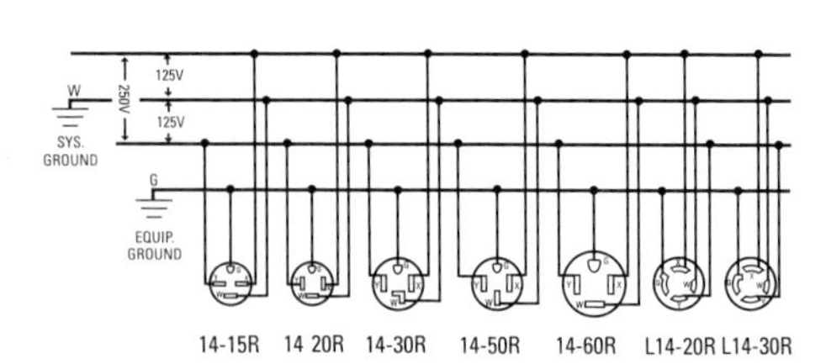

Courtesy of

WIRING DIAGRAMS FOR NEMA CONFIGURATIONS

3 Pole, 4 Wire Grounding 3ø 250V

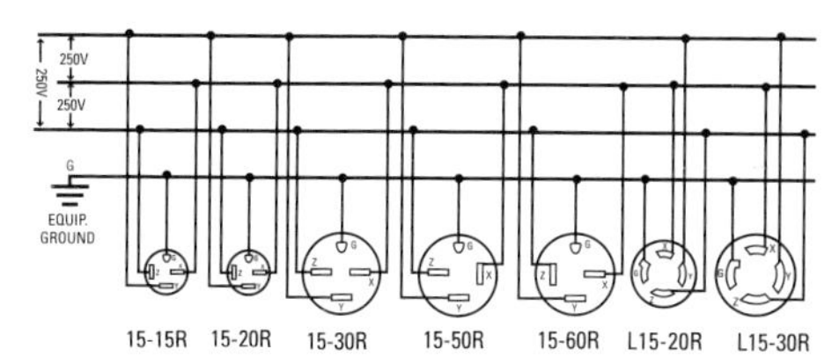

3 Pole, 4 Wire Grounding 3ø 480V

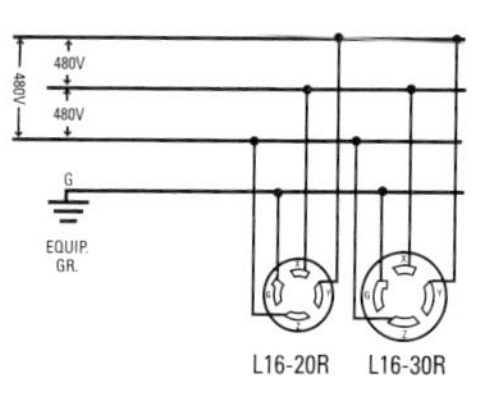

3 Pole, 4 Wire Grounding 3ø 600V

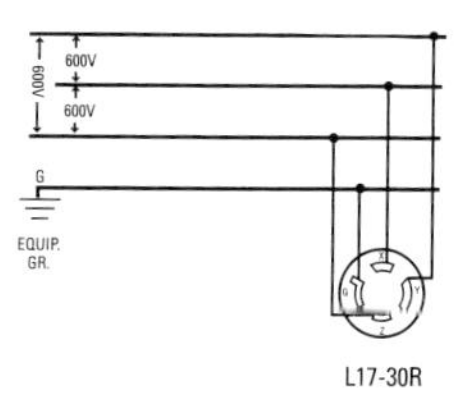

Courtesy of

COOPER Wiring Devices

The New Power in Wiring Devices

WIRING DIAGRAMS FOR NEMA CONFIGURATIONS

4 Pole, 4 Wire
Nongrounding
3ø 120/208V

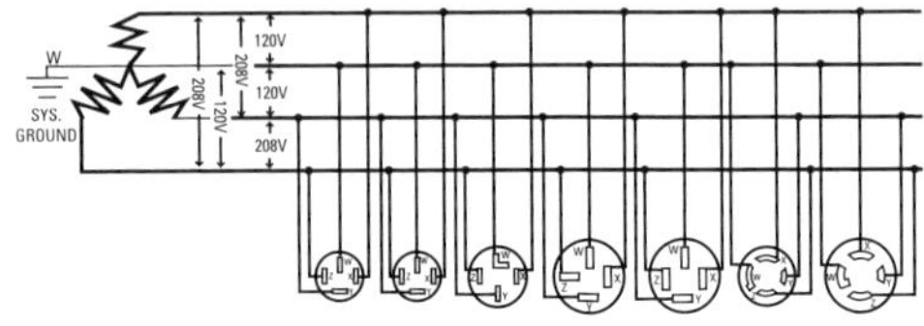

4 Pole, 4 Wire
Nongrounding
3ø 277/480V

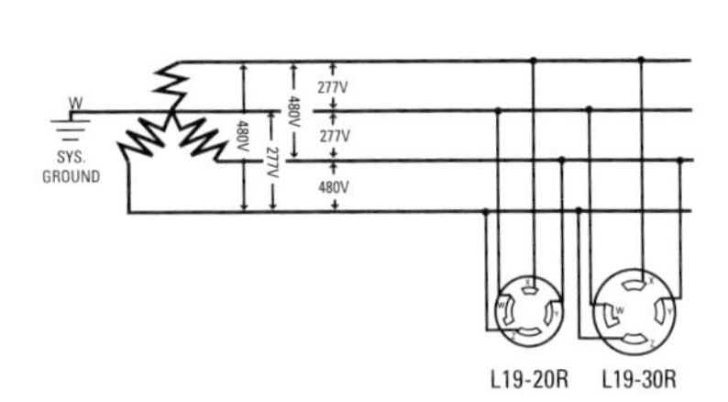

4 Pole, 4 Wire
Nongrounding
3ø 347/600V

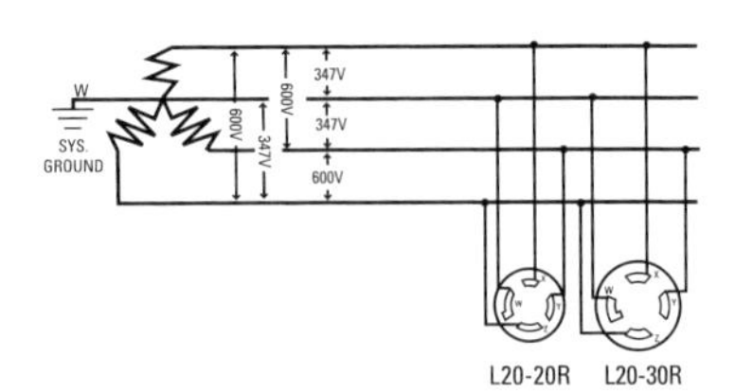

Courtesy of

COOPER Wiring Devices
The New Power in Wiring Devices

WIRING DIAGRAMS FOR NEMA CONFIGURATIONS

4 Pole,5 Wire Grounding 3ø 120/208V

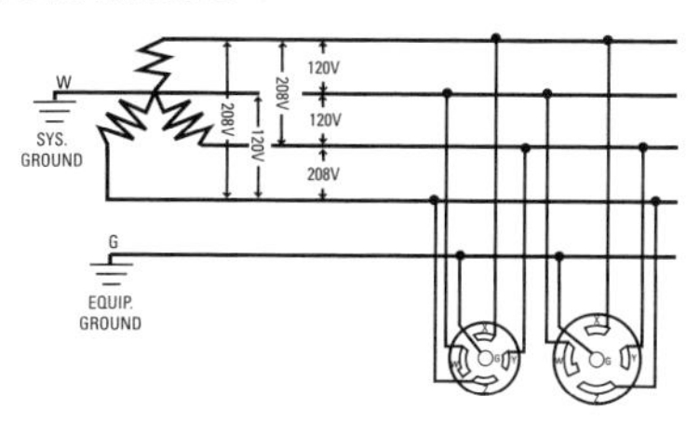

L21-20R L21-30R

4 Pole,5 Wire Grounding 3ø 277/480V

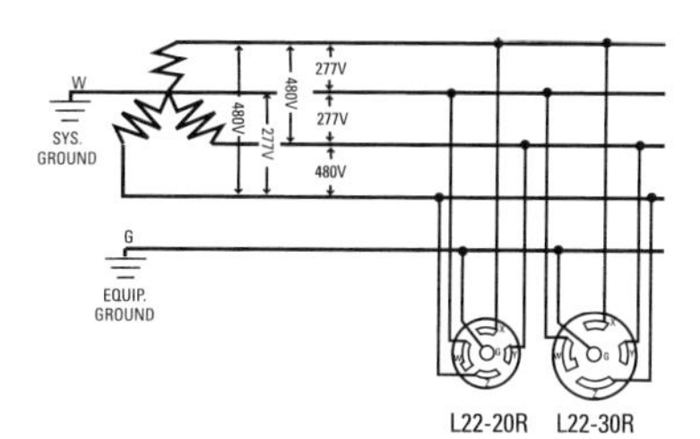

L22-20R L22-30R

4 Pole,5 Wire Grounding 3ø 347/600V

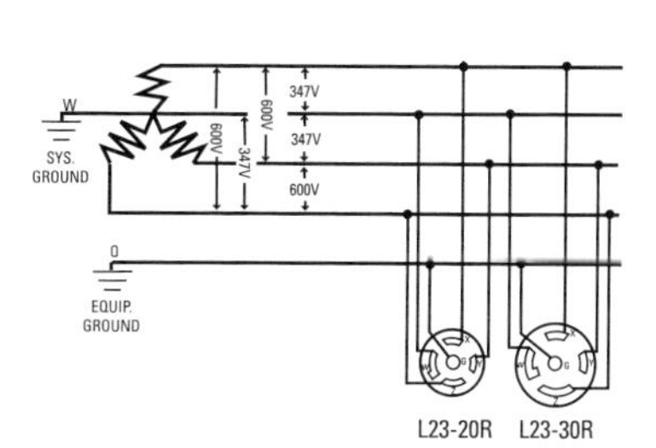

L23-20R L23-30R

NEMA ENCLOSURE TYPES

The specific enclosure types, their applications, and the environmental conditions they are designed to provide a degree of protection against are as follows:

Type 1 – Enclosures constructed for indoor use to provide a degree of protection to personnel against access to hazardous parts and to provide a degree of protection of the equipment inside the enclosure against ingress of solid foreign objects (falling dirt).

Type 2 – Enclosures constructed for indoor use to provide a degree of protection to personnel against access to hazardous parts; to provide a degree of protection of the equipment inside the enclosure against ingress of solid foreign objects (falling dirt); and to provide a degree of protection with respect to harmful effects on the equipment due to the ingress of water (dripping and light splashing).

Type 3 – Enclosures constructed for either indoor or outdoor use to provide a degree of protection to personnel against access to hazardous parts; to provide a degree of protection of the equipment inside the enclosure against ingress of solid foreign objects (falling dirt and windblown dust); to provide a degree of protection with respect to harmful effects on the equipment due to the ingress of water (rain, sleet, snow); and that will be undamaged by the external formation of ice on the enclosure.

Type 3R – Enclosures constructed for either indoor or outdoor use to provide a degree of protection to personnel against access to hazardous parts; to provide a degree of protection of the equipment inside the enclosure against ingress of solid foreign objects (falling dirt); to provide a degree of protection with respect to harmful effects on the equipment due to the ingress of water (rain, sleet, snow); and that will be undamaged by the external formation of ice on the enclosure.

NEMA ENCLOSURE TYPES

Type 3S – Enclosures constructed for either indoor or outdoor use to provide a degree of protection to personnel against access to hazardous parts; to provide a degree of protection of the equipment inside the enclosure against ingress of solid foreign objects (falling dirt and windblown dust); to provide a degree of protection with respect to harmful effects on the equipment due to the ingress of water (rain, sleet, snow); and for which the external mechanism(s) remain operable when ice laden.

Type 3X – Enclosures constructed for either indoor or outdoor use to provide a degree of protection to personnel against access to hazardous parts; to provide a degree of protection of the equipment inside the enclosure against ingress of solid foreign objects (falling dirt and windblown dust); to provide a degree of protection with respect to harmful effects on the equipment due to the ingress of water (rain, sleet, snow); that provides an increased level of protection against corrosion and that will be undamaged by the external formation of ice on the enclosure.

Type 3RX – Enclosures constructed for either indoor or outdoor use to provide a degree of protection to personnel against access to hazardous parts; to provide a degree of protection of the equipment inside the enclosure against ingress of solid foreign objects (falling dirt); to provide a degree of protection with respect to harmful effects on the equipment due to the ingress of water (rain, sleet, snow); that will be undamaged by the external formation of ice on the enclosure that provides an increased level of protection against corrosion.

Type 3SX – Enclosures constructed for either indoor or outdoor use to provide a degree of protection to personnel against access to hazardous parts; to provide a degree of protection of the equipment inside the enclosure against ingress of solid foreign objects (falling dirt and windblown dust); to provide a degree of protection with respect to harmful effects on the equipment due to the ingress of water (rain, sleet, snow); that provides an increased level of protection against corrosion; and for which the external mechanism(s) remain operable when ice laden.

Type 4 – Enclosures constructed for either indoor or outdoor use to provide a degree of protection to personnel against access to hazardous parts; to provide a degree of protection of the equipment inside the enclosure against ingress of solid foreign objects (falling dirt and windblown dust); to provide a degree of protection with respect to harmful effects on the equipment due to the ingress of water (rain, sleet, snow, splashing water, and hose-directed water); and that will be undamaged by the external formation of ice on the enclosure.

Type 4X – Enclosures constructed for either indoor or outdoor use to provide a degree of protection to personnel against access to hazardous parts; to provide a degree of protection of the equipment inside the enclosure against ingress of solid foreign objects (falling dirt and windblown dust); to provide a degree of protection with respect to harmful effects on the equipment due to the ingress of water (rain, sleet, snow, splashing water, and hose-directed water); that provides an increased level of protection against corrosion; and that will be undamaged by the external formation of ice on the enclosure.

Type 5 – Enclosures constructed for indoor use to provide a degree of protection to personnel against access to hazardous parts; to provide a degree of protection of the equipment inside the enclosure against ingress of solid foreign objects (falling dirt and settling airborne dust, lint, fibers, and flyings); and to provide a degree of protection with respect to harmful effects on the equipment due to the ingress of water (dripping and light splashing).

Type 6 – Enclosures constructed for either indoor or outdoor use to provide a degree of protection to personnel against access to hazardous parts; to provide a degree of protection of the equipment inside the enclosure against ingress of solid foreign objects (falling dirt); to provide a degree of protection with respect to harmful effects on the equipment due to the ingress of water (hose-directed water and the entry of water during occasional temporary submersion at a limited depth); and that will be undamaged by the external formation of ice on the enclosure.

Type 6P – Enclosures constructed for either indoor or outdoor use to provide a degree of protection to personnel against access to hazardous parts; to provide a degree of protection of the equipment inside the enclosure against ingress of solid foreign objects (falling dirt); to provide a degree of protection with respect to harmful effects on the equipment due to the ingress of water (hose-directed water and the entry of water during prolonged submersion at a limited depth); that provides an increased level of protection against corrosion; and that will be undamaged by the external formation of ice on the enclosure.

Type 12 – Enclosures constructed (without knockouts) for indoor use to provide a degree of protection to personnel against access to hazardous parts; to provide a degree of protection of the equipment inside the enclosure against ingress of solid foreign objects (falling dirt and circulating dust, lint, fibers, and flyings); to provide a degree of protection with respect to harmful effects on the equipment due to the ingress of water (dripping and light splashing); and to provide a degree of protection against light splashing and seepage of oil and non-corrosive coolants.

Type 12K – Enclosures constructed (with knockouts) for indoor use to provide a degree of protection to personnel against access to hazardous parts; to provide a degree of protection of the equipment inside the enclosure against ingress of solid foreign objects (falling dirt and circulating dust, lint, fibers, and flyings); to provide a degree of protection with respect to harmful effects on the equipment due to the ingress of water (dripping and light splashing); and to provide a degree of protection against light splashing and seepage of oil and non-corrosive coolants.

Type 13 – Enclosures constructed for indoor use to provide a degree of protection to personnel against access to hazardous parts; to provide a degree of protection of the equipment inside the enclosure against ingress of solid foreign objects (falling dirt and circulating dust, lint, fibers, and flyings); to provide a degree of protection with respect to harmful effects on the equipment due to the ingress of water (dripping and light splashing); and to provide a degree of protection against the spraying, splashing, and seepage of oil and non-corrosive coolants.

U.S. WEIGHTS AND MEASURES

Linear Measures

	1 INCH	= 2.540 CENTIMETERS
12 INCHES	= 1 FOOT	= 3.048 DECIMETERS
3 FEET	= 1 YARD	= 9.144 DECIMETERS
5.5 YARDS	= 1 ROD	= 5.029 METERS
40 RODS	= 1 FURLONG	= 2.018 HECTOMETERS
8 FURLONGS	= 1 MILE	= 1.609 KILOMETERS

Mile Measurements

1 STATUTE MILE	=	5,280 FEET
1 SCOTS MILE	=	5,952 FEET
1 IRISH MILE	=	6,720 FEET
1 RUSSIAN VERST	=	3,504 FEET
1 ITALIAN MILE	=	4,401 FEET
1 SPANISH MILE	=	15,084 FEET

Other Linear Measurements

1 HAND	=	4 INCHES
1 SPAN	=	9 INCHES
1 CHAIN	=	22 YARDS
1 LINK	=	7.92 INCHES
1 FATHOM	=	6 FEET
1 FURLONG	=	10 CHAINS
1 CABLE	=	608 FEET

Square Measures

144	SQUARE INCHES	=	1	SQUARE FOOT
9	SQUARE FEET	=	1	SQUARE YARD
30¼	SQUARE YARDS	=	1	SQUARE ROD
40	RODS	=	1	ROOD
4	ROODS	=	1	ACRE
640	ACRES	=	1	SQUARE MILE
1	SQUARE MILE	=	1	SECTION
36	SECTIONS	=	1	TOWNSHIP

Cubic or Solid Measures

1	CU. FOOT	=	1728	CU. INCHES
1	CU. YARD	=	27	CU. FEET
1	CU. FOOT	=	7.48	GALLONS
1	GALLON (WATER)	=	8.34	LBS.
1	GALLON (U.S.)	=	231	CU. INCHES OF WATER
1	GALLON (IMPERIAL)	=	277¼	CU. INCHES OF WATER

U.S. WEIGHTS AND MEASURES

Liquid Measurements

1 PINT	=	4 GILLS
1 QUART	=	2 PINTS
1 GALLON	=	4 QUARTS
1 FIRKIN	=	9 GALLONS (ALE OR BEER)
1 BARREL	=	42 GALLONS (PETROLEUM OR CRUDE OIL)

Dry Measure

1 QUART	=	2 PINTS
1 PECK	=	8 QUARTS
1 BUSHEL	=	4 PECKS

Weight Measurement (Mass)

A. Avoirdupois Weight:

1 OUNCE	=	16	DRAMS
1 POUND	=	16	OUNCES
1 HUNDREDWEIGHT	=	100	POUNDS
1 TON	=	2000	POUNDS

B. Troy Weight:

1 CARAT	=	3.17	GRAINS
1 PENNYWEIGHT	=	20	GRAINS
1 OUNCE	=	20	PENNYWEIGHTS
1 POUND	=	12	OUNCES
1 LONG HUNDRED-WEIGHT	=	112	POUNDS
1 LONG TON	=	20	LONG HUNDREDWEIGHTS
	=	2240	POUNDS

C. Apothecaries Weight:

1 SCRUPLE	=	20	GRAINS	=	1.296	GRAMS
1 DRAM	=	3	SCRUPLES	=	3.888	GRAMS
1 OUNCE	=	8	DRAMS	=	31.1035	GRAMS
1 POUND	=	12	OUNCES	=	373.2420	GRAMS

D. Kitchen Weights and Measures:

1 U.S. PINT	=	16	FL. OUNCES
1 STANDARD CUP	=	8	FL. OUNCES
1 TABLESPOON	=	0.5	FL. OUNCES (15 CU. CMs.)
1 TEASPOON	=	0.16	FL. OUNCES (5 CU. CMs.)

METRIC SYSTEM

Prefixes

A. MEGA	= 1,000,000	E. DECI	= 0.1
B. KILO	= 1000	F. CENTI	= 0.01
C. HECTO	= 100	G. MILLI	= 0.001
D. DEKA	= 10	H. MICRO	= 0.000001

Linear Measure

(THE UNIT IS THE METER = 39.37 INCHES)

1 CENTIMETER	= 10	MILLIMETERS	=	0.3937011	IN.
1 DECIMETER	= 10	CENTIMETERS	=	3.9370113	INs.
1 METER	= 10	DECIMETERS	=	1.0936143	YDs.
			=	3.2808429	FT.
1 DEKAMETER	= 10	METERS	=	10.936143	YDs.
1 HECTOMETER	= 10	DEKAMETERS	=	109.36143	YDs.
1 KILOMETER	= 10	HECTOMETERS	=	0.62137	MILE
1 MYRIAMETER	= 10,000	METERS			

Square Measure

(THE UNIT IS THE SQUARE METER = 1549.9969 SQ. INCHES)

1 SQ. CENTIMETER	= 100 SQ. MILLIMETERS	=	0.1550	SQ. IN.
1 SQ. DECIMETER	= 100 SQ. CENTIMETERS	=	15.550	SQ. INs.
1 SQ. METER	= 100 SQ. DECIMETERS	=	10.7639	SQ. FT.
1 SQ. DEKAMETER	= 100 SQ. METERS	=	119.60	SQ. YDs.
1 SQ. HECTOMETER	= 100 SQ. DEKAMETERS			
1 SQ. KILOMETER	= 100 SQ. HECTOMETERS			

(THE UNIT IS THE "ARE" = 100 SQ. METERS)

1 CENTIARE	= 10	MILLIARES	=	10.7643	SQ. FT.
1 DECIARE	= 10	CENTIARES	=	11.96033	SQ. YDs.
1 ARE	= 10	DECIARES	=	119.6033	SQ. YDs.
1 DEKARE	= 10	ARES	=	0.247110	ACRES
1 HEKTARE	= 10	DEKARES	=	2.471098	ACRES
1 SQ. KILOMETER	= 100	HEKTARES	=	0.38611	SQ. MILE

Cubic Measure

(THE UNIT IS THE "STERE" = 61,025.38659 CU. INs.)

1 DECISTERE	= 10 CENTISTERES	=	3.531562	CU. FT.
1 STERE	= 10 DECISTERES	=	1.307986	CU. YDs.
1 DEKASTERE	= 10 STERES	=	13.07986	CU. YDs.

METRIC SYSTEM

Cubic Measure

(THE UNIT IS THE METER = 39.37 INCHES)

1 CU. CENTIMETER	= 1000 CU. MILLIMETERS	=	0.06102 CU. IN.
1 CU. DECIMETER	= 1000 CU. CENTIMETERS	=	61.02374 CU. IN.
1 CU. METER	= 1000 CU. DECIMETERS	=	35.31467 CU. FT.
	= 1 STERE	=	1.30795 CU. YDs.
1 CU. CENTIMETER (WATER)		=	1 GRAM
1000 CU. CENTIMETERS (WATER) = 1 LITER		=	1 KILOGRAM
1 CU. METER (1000 LITERS)		=	1 METRIC TON

Measures of Weight

(THE UNIT IS THE GRAM = 0.035274 OUNCES)

1 MILLIGRAM	=		=	0.015432	GRAINS
1 CENTIGRAM	=	10 MILLIGRAMS	=	0.15432	GRAINS
1 DECIGRAM	=	10 CENTIGRAMS	=	1.5432	GRAINS
1 GRAM	=	10 DECIGRAMS	=	15.4323	GRAINS
1 DEKAGRAM	=	10 GRAMS	=	5.6438	DRAMS
1 HECTOGRAM	=	10 DEKAGRAMS	=	3.5274	OUNCES
1 KILOGRAM	=	10 HECTOGRAMS	=	2.2046223	POUNDS
1 MYRIAGRAM	=	10 KILOGRAMS	=	22.046223	POUNDS
1 QUINTAL	=	10 MYRIAGRAMS	=	1.986412	CWT.
1 METRIC TON	=	10 QUINTAL	=	22,045.622	POUNDS
1 GRAM	=	0.56438 DRAMS			
1 DRAM	=	1.77186 GRAMS			
	=	27.3438 GRAINS			
1 METRIC TON	=	2204.6223 POUNDS			

Measures of Capacity

(THE UNIT IS THE LITER = 1.0567 LIQUID QUARTS)

1 CENTILITER	=	10 MILLILITERS	=	0.338	FLUID OUNCES
1 DECILITER	=	10 CENTILITERS	=	3.38	FLUID OUNCES
1 LITER	=	10 DECILITERS	=	33.8	FLUID OUNCES
1 DEKALITER	=	10 LITERS	=	0.284	BUSHEL
1 HECTOLITER	=	10 DEKALITERS	=	2.84	BUSHELS
1 KILOLITER	=	10 HECTOLITERS	=	264.2	GALLONS

NOTE: $\frac{\text{KILOMETERS}}{8} \times 5 = \text{MILES}$ OR $\frac{\text{MILES}}{5} \times 8 = \text{KILOMETERS}$

METRIC SYSTEM

Metric Designator and Trade Sizes

METRIC DESIGNATOR												
12	16	21	27	35	41	53	63	78	91	103	129	155
3/8	1/2	3/4	1	1 1/4	1 1/2	2	2 1/2	3	3 1/2	4	5	6
TRADE SIZE												

Source: NFPA 70, *National Electrical Code®*, NFPA, Quincy, MA, 2010, Table 300.1, as modified.

U.S. Weights and Measures/Metric Equivalent Chart

	In.	Ft.	Yd.	Mile	Mm	Cm	M	Km
1 Inch =	1	.0833	.0278	1.578×10^{-5}	25.4	**2.54**	.0254	2.54×10^{-5}
1 Foot =	12	1	.333	1.894×10^{-4}	304.8	**30.48**	.3048	3.048×10^{-4}
1 Yard =	36	3	1	5.6818×10^{-4}	914.4	91.44	**.9144**	9.144×10^{-4}
1 Mile =	63,360	5280	1,760	1	1,609,344	160,934.4	1609.344	**1,609,344**
1 mm =	**.03937**	.00032808	1.0936×10^{-3}	6.2137×10^{-7}	1	0.1	0.001	0.000001
1 cm =	**.3937**	.0328084	.0109361	6.2137×10^{-6}	10	1	0.01	0.00001
1 m =	39.37	3.280.84	**1.093.61**	6.2137×10^{-4}	1000	100	1	0.001
1 km =	39,370	3,280.84	1,093.61	**0.62137**	1,000,000	100,000	1000	1

In. = Inches Ft. = Foot Yd. = Yard Mm = Millimeter Cm = Centimeter M = Meter Km = Kilometer

Explanation of Scientific Notation

Scientific notation is simply a way of expressing very large or very small numbers in a more compact format. Any number can be expressed as a number between 1 and 10, multiplied by a power of 10 (which indicates the correct position of the decimal point in the original number). Numbers greater than 10 have positive powers of 10, and numbers less than 1 have negative powers of 10.

Example: $186{,}000 = 1.86 \times 10^5$ $0.000524 = 5.24 \times 10^{-4}$

Useful Conversions/Equivalents

1 BTU . . . . Raises 1 lb. of water 1°F
1 GRAM CALORIE . . . . Raises 1 gram of water 1°C
1 CIRCULAR MIL . . . . Equals 0.7854 sq. mil
1 SQ. MIL . . . . Equals 1.27 cir. mils
1 MIL . . . . Equals 0.001 in.

To determine circular mil of a conductor:

ROUND CONDUCTOR . . . . $CM = (\text{Diameter in mils})^2$

BUS BAR . . . . $CM = \dfrac{\text{Width (mils)} \times \text{Thickness (mils)}}{0.7854}$

Notes: 1 millimeter = 39.37 mils 1 cir. millimeter = 1550 cir. mils
1 sq. millimeter = 1974 cir. mils

DECIMAL EQUIVALENTS

FRACTION					DECIMAL	FRACTION					DECIMAL
1/64					.0156	33/64					.5156
2/64	1/32				.0313	34/64	17/32				.5313
3/64					.0469	35/64					.5469
4/64	2/32	1/16			.0625	36/64	18/32	9/16			.5625
5/64					.0781	37/64					.5781
6/64	3/32				.0938	38/64	19/32				.5938
7/64					.1094	39/64					.6094
8/64	4/32	2/16	1/8		.125	40/64	20/32	10/16	5/8		.625
9/64					.1406	41/64					.6406
10/64	5/32				.1563	42/64	21/32				.6563
11/64					.1719	43/64					.6719
12/64	6/32	3/16			.1875	44/64	22/32	11/16			.6875
13/64					.2031	45/64					.7031
14/64	7/32				.2188	46/64	23/32				.7188
15/64					.2344	47/64					.7344
16/64	8/32	4/16	2/8	1/4	.25	48/64	24/32	12/16	6/8	3/4	.75
17/64					.2656	49/64					.7656
18/64	9/32				.2813	50/64	25/32				.7813
19/64					.2969	51/64					.7969
20/64	10/32	5/16			.3125	52/64	26/32	13/16			.8125
21/64					.3281	53/64					.8281
22/64	11/32				.3438	54/64	27/32				.8438
23/64					.3594	55/64					.8594
24/64	12/32	6/16	3/8		.375	56/64	28/32	14/16	7/8		.875
25/64					.3906	57/64					.8906
26/64	13/32				.4063	58/64	29/32				.9063
27/64					.4219	59/64					.9219
28/64	14/32	7/16			.4375	60/64	30/32	15/16			.9375
29/64					.4531	61/64					.9531
30/64	15/32				.4688	62/64	31/32				.9688
31/64					.4844	63/64					.9844
32/64	16/32	8/16	4/8	2/4	.5	64/64	32/32	16/16	8/8	4/4	1.000

Decimals are rounded to the nearest 10,000th.

TWO-WAY CONVERSION TABLE

To convert from the unit of measure in Column B to the unit of measure in Column C, multiply the number of units in Column B by the multiplier in Column A. To convert from Column C to B, use the multiplier in Column D.

EXAMPLE: To convert 1000 BTUs to CALORIES, find the "BTU - CALORIE" combination in Columns B and C. "BTU" is in Column B and "CALORIE" is in Column C; so we are converting from B to C. Therefore, we use Column A multiplier. 1000 BTUs x 251.996 = 251,996 Calories.

To convert 251,996 Calories to BTUs, use the same "BTU - CALORIE" combination. But this time you are converting from C to B. Therefore, use Column D multiplier. 251,996 Calories x .0039683 = 1000 BTUs.

A x B = C & **D x C = B**

To convert from B to C, Multiply B x A:

To convert from C to B, Multiply C x D:

A	B	C	D
43,560	**Acre**	**Sq. Foot**	2.2956×10^{-5}
1.5625×10^{-3}	**Acre**	**Sq. Mile**	640
6.4516	**Ampere per sq. cm.**	**Ampere per sq. in.**	0.155003
1.256637	**Ampere (turn)**	**Gilberts**	0.79578
33.89854	**Atmosphere**	**Foot of H_2O**	0.029499
29.92125	**Atmosphere**	**Inch of Hg**	0.033421
14.69595	**Atmosphere**	**Pound force/sq. in.**	0.06804
251.996	**BTU**	**Calorie**	3.96832×10^{-3}
778.169	**BTU**	**Foot-pound force**	1.28507×10^{-3}
3.93015×10^{-4}	**BTU**	**Horsepower-hour**	2544.43
1055.056	**BTU**	**Joule**	9.47817×10^{-4}
2.9307×10^{-4}	**BTU**	**Kilowatt-hour**	3412.14
3.93015×10^{-4}	**BTU/hour**	**Horsepower**	2544.43
2.93071×10^{-4}	**BTU/hour**	**Kilowatt**	3412.1412
0.293071	**BTU/hour**	**Watt**	3.41214
4.19993	**BTU/minute**	**Calorie/second**	0.23809
0.0235809	**BTU/minute**	**Horsepower**	42.4072
17.5843	**BTU/minute**	**Watt**	0.0568

(continues)

TWO-WAY CONVERSION TABLE

To convert from B to C, Multiply B x A:

To convert from C to B, Multiply C x D:

A	B	C	D
4.1868	Calorie	Joule	0.238846
0.0328084	Centimeter	Foot	30.48
0.3937	Centimeter	Inch	2.54
0.00001	Centimeter	Kilometer	100,000
0.01	Centimeter	Meter	100
6.2137×10^{-6}	Centimeter	Mile	160,934.4
10	Centimeter	Millimeter	0.1
0.010936	Centimeter	Yard	91.44
$7.85398 \text{ X } 10^{-7}$	Circular mil	Sq. Inch	1.273239×10^{6}
0.000507	Circular mil	Sq. Millimeter	1973.525
0.06102374	Cubic Centimeter	Cubic Inch	16.387065
0.028317	Cubic Foot	Cubic Meter	35.31467
1.0197×10^{-3}	Dyne	Gram Force	980.665
1×10^{-5}	Dyne	Newton	100,000
1	Dyne centimeter	Erg	1
7.376×10^{-8}	Erg	Foot pound force	1.355818×10^{7}
2.777×10^{-14}	Erg	Kilowatt-hour	3.6×10^{13}
1.0×10^{-7}	Erg/second	Watt	1.0×10^{7}
12	Foot	Inch	0.0833
3.048×10^{-4}	Foot	Kilometer	3280.84
0.3048	Foot	Meter	3.28084
1.894×10^{-4}	Foot	Mile	5280
304.8	Foot	Millimeter	0.00328
0.333	Foot	Yard	3
10.7639	Foot candle	Lux	0.0929
0.882671	Foot of H_2O	Inch of Hg	1.13292
5.0505×10^{-7}	Foot pound force	Horsepower-hour	$1.98 \text{ X } 10^{6}$
1.35582	Foot pound force	Joule	0.737562
3.76616×10^{-7}	Foot pound force	Kilowatt-hour	2.655223×10^{6}
3.76616×10^{-4}	Foot pound force	Watt-hour	2655.22
3.76616×10^{-7}	Foot pnd. force/hour	Kilowatt	2.6552×10^{6}
3.0303×10^{-5}	Foot pnd. force/minute	Horsepower	33,000

TWO-WAY CONVERSION TABLE

To convert from B to C, Multiply B x A:

To convert from C to B, Multiply C x D:

A	B	C	D
2.2597×10^{-5}	Foot pnd. force/minute	Kilowatt	44,253.7
0.022597	Foot pnd. force/minute	Watt	44.2537
1.81818×10^{-3}	Foot pnd. force/second	Horsepower	550
1.355818×10^{-3}	Foot pnd. force/second	Kilowatt	737.562
0.7457	Horsepower	Kilowatt	1.34102
745.7	Horsepower	Watt	0.00134
0.0022046	Gram	Pound mass	453.592
2.54×10^{-5}	Inch	Kilometer	39,370
0.0254	Inch	Meter	39.37
1.578×10^{-5}	Inch	Mile	63,360
25.4	Inch	Millimeter	0.03937
0.0278	Inch	Yard	36
0.07355	Inch of H_2O	Inch of Hg	13.5951
2.7777×10^{-7}	Joule	Kilowatt-hour	3.6×10^{6}
2.7777×10^{-4}	Joule	Watt hour	3600
1	Joule	Watt second	1
1000	Kilometer	Meter	0.001
0.62137	Kilometer	Mile	1.609344
1,000,000	Kilometer	Millimeter	0.000001
1093.61	Kilometer	Yard	9.144×10^{-4}
0.000621	Meter	Mile	1609.344
1000	Meter	Millimeter	0.001
1.0936	Meter	Yard	0.9144
1,609,344	Mile	Millimeter	6.2137×10^{-7}
1760	Mile	Yard	5.681×10^{-4}
1.0936×10^{-3}	Millimeter	Yard	914.4
0.224809	Newton	Pound force	4.44822
0.03108	Pound	Slug	32.174
0.0005	Pound	Ton (short)	2000
0.155	Sq. Centimeter	Sq. Inch	6.4516
0.092903	Sq. Foot	Sq. Meter	10.76391
0.386102	Sq. Kilometer	Sq. Mile	2.589988

METALS

METAL	SYMB	SPEC. GRAV.	MELT POINT C°	MELT POINT F°	ELEC. COND. % COPPER	LBS. CU."
ALUMINUM	AL	2.71	660	1220	64.9	.0978
ANTIMONY	SB	6.62	630	1167	4.42	.2390
ARSENIC	AS	5.73	–	–	4.9	.2070
BERYLLIUM	BE	1.83	1280	2336	9.32	.0660
BISMUTH	BI	9.80	271	520	1.50	.3540
BRASS (70-30)		8.51	900	1652	28.0	.3070
BRONZE (5% SN)		8.87	1000	1832	18.0	.3200
CADMIUM	CD	8.65	321	610	22.7	.3120
CALCIUM	CA	1.55	850	1562	50.1	.0560
COBALT	CO	8.90	1495	2723	17.8	.3210
COPPER	CU					
ROLLED		8.89	1083	1981	100.0	.3210
TUBING		8.95	–	–	100.0	.3230
GOLD	AU	19.30	1063	1945	71.2	.6970
GRAPHITE		2.25	3500	6332	10^{-3}	.0812
INDIUM	IN	7.30	156	311	20.6	.2640
IRIDIUM	IR	22.40	2450	4442	32.5	.8090
IRON	FE	7.20	1200–1400	2192–2552	17.6	.2600
MALLEABLE		7.20	1500–1600	2732–2912	10	.2600
WROUGHT		7.70	1500–1600	2732–2912	10	.2780
LEAD	PB	11.40	327	621	8.35	.4120
MAGNESIUM	MG	1.74	651	1204	38.7	.0628

METALS

METAL	SYMB	SPEC. GRAV.	MELT POINT C°	F°	ELEC. COND. % COPPER	LBS. CU."
MANGANESE	MN	7.20	1245	2273	0.9	.2600
MERCURY	HG	13.65	–38.9	–37.7	1.80	.4930
MOLYBDENUM	MO	10.20	2620	4748	36.1	.3680
MONEL (63 - 37)		8.87	1300	2372	3.0	.3200
NICKEL	NI	8.90	1452	2646	25.0	.3210
PHOSPHOROUS	P	1.82	44.1	111.4	10^{-17}	.0657
PLATINUM	PT	21.46	1773	3221	17.5	.7750
POTASSIUM	K	0.860	62.3	144.1	28	.0310
SELENIUM	SE	4.81	220	428	14.4	.1740
SILICON	SI	2.40	1420	2588	10^{-5}	.0866
SILVER	AG	10.50	960	1760	106	.3790
STEEL (CARBON)		7.84	1330–1380	2436–2516	10	.2830
STAINLESS						
(18-8)		7.92	1500	2732	2.5	.2860
(13-CR)		7.78	1520	2768	3.5	.2810
TANTALUM	TA	16.60	2900	5414	13.9	.599
TELLURIUM	TE	6.20	450	846	10^{-5}	.224
THORIUM	TH	11.70	1845	3353	9.10	.422
TIN	SN	7.30	232	449	15.00	.264
TITANIUM	TI	4.50	1800	3272	2.10	.162
TUNGSTEN	W	19.30	3410	–	31.50	.697
URANIUM	U	18.70	1130	2066	2.80	.675
VANADIUM	V	5.96	1710	3110	6.63	.215
ZINC	ZN	7.14	410	786	29.10	.258
ZIRCONIUM	ZR	6.40	1700	3092	4.20	.231

METALS

Specific Resistance (K)

The specific resistance (K) of a material is the resistance offered by a wire of this material that is 1 foot long with a diameter of 1 mil.

MATERIAL	"K"	MATERIAL	"K"
BRASS	43.0	ALUMINUM	17.0
CONSTANTAN	295	MONEL	253
COPPER	10.8	NICHROME	600
GERMAN SILVER 18%	200	NICKEL	947
GOLD	14.7	TANTALUM	93.3
IRON (PURE)	60.0	TIN	69.0
MAGNESIUM	276	TUNGSTEN	34.0
MANGANIN	265	SILVER	9.7

NOTE: 1. The resistance of a wire is directly proportional to the specific resistance of the material.

2. "K" = Specific Resistance

3. Resistance varies with temperature. See *NEC* Chapter 9, Table 8, Notes.

CENTIGRADE AND FAHRENHEIT THERMOMETER SCALES

DEG-C	DEG-F	DEG-C	DEG-F	DEG-C	DEG-F	DEG-C	DEG-F
0	32						
1	33.8	26	78.8	51	123.8	76	168.8
2	35.6	27	80.6	52	125.6	77	170.6
3	37.4	28	82.4	53	127.4	78	172.4
4	39.2	29	84.2	54	129.2	79	174.2
5	41	30	86	55	131	80	176
6	42.8	31	87.8	56	132.8	81	177.8
7	44.6	32	89.6	57	134.6	82	179.6
8	46.4	33	91.4	58	136.4	83	181.4
9	48.2	34	93.2	59	138.2	84	183.2
10	50	35	95	60	140	85	185
11	51.8	36	96.8	61	141.8	86	186.8
12	53.6	37	98.6	62	143.6	87	188.6
13	55.4	38	100.4	63	145.4	88	190.4
14	57.2	39	102.2	64	147.2	89	192.2
15	59	40	104	65	149	90	194
16	60.8	41	105.8	66	150.8	91	195.8
17	62.6	42	107.6	67	152.6	92	197.6
18	64.4	43	109.4	68	154.4	93	199.4
19	66.2	44	111.2	69	156.2	94	201.2
20	68	45	113	70	158	95	203
21	69.8	46	114.8	71	159.8	96	204.8
22	71.6	47	116.6	72	161.6	97	206.6
23	73.4	48	118.4	73	163.4	98	208.4
24	75.2	49	120.2	74	165.2	99	210.2
25	77	50	122	75	167	100	212

- TEMP. C° = $^5/_9$ x (TEMP. F° – 32)
- TEMP. F° = ($^9/_5$ x TEMP. C°) + 32
- Ambient temperature is the temperature of the surrounding cooling medium.
- Rated temperature rise is the permissible rise in temperature above ambient when operating under load.

USEFUL MATH FORMULAS

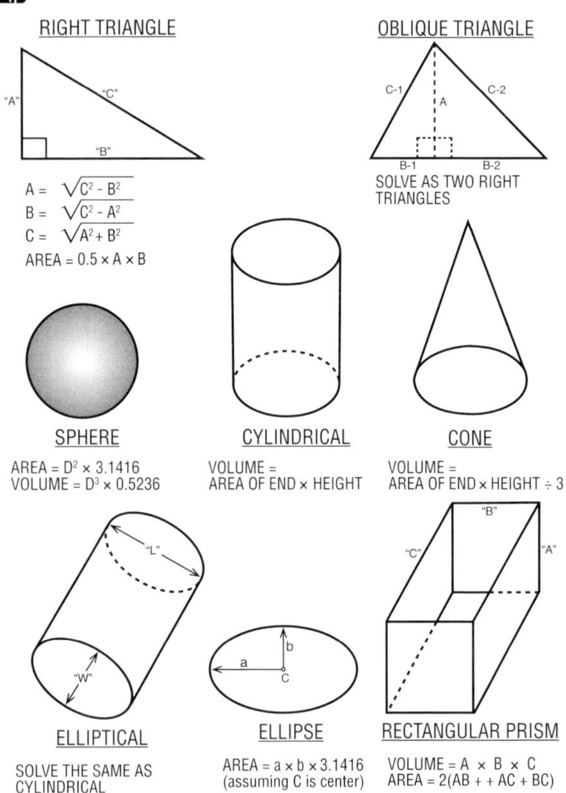

THE CIRCLE

DEFINITION: A closed plane curve having every point an equal distance from a fixed point within the curve

CIRCUMFERENCE:	The distance around a circle
DIAMETER:	The distance across a circle through the center
RADIUS:	The distance from the center to the edge of a circle
ARC:	A part of the circumference
CHORD:	A straight line connecting the ends of an arc
SEGMENT:	An area bounded by an arc and a chord
SECTOR:	A part of a circle enclosed by two radii and the arc that they cut off

CIRCUMFERENCE OF A CIRCLE = 3.1416 x 2 x radius

AREA OF A CIRCLE = 3.1416 x radius^2

ARC LENGTH = Degrees in arc x radius x 0.01745

RADIUS LENGTH = One-half length of diameter

SECTOR AREA = One-half length of arc x radius

CHORD LENGTH = $2\sqrt{A \times B}$

SEGMENT AREA = Sector area minus triangle area

NOTE:

3.1416 x 2 x R = 360 Degrees, or 0.0087266 x 2 x R = 1 Degree, or 0.01745 x R = 1 Degree

This gives us the arc formula.

DEGREES x RADIUS x 0.01745 = DEVELOPED LENGTH

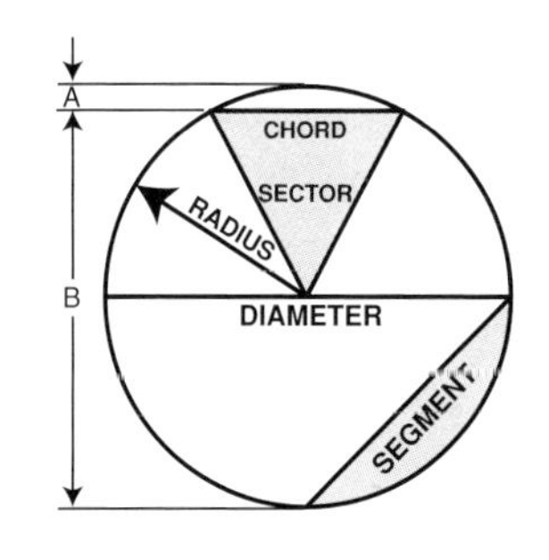

EXAMPLE:

For a ninety degree conduit bend, having a radius of 17.25":

90 x 17.25" x 0.01745 = Developed Length

27.09" = Developed Length

FRACTIONS

Definitions

A. A FRACTION is a quantity less than a unit.

B. A NUMERATOR is the term of a fraction indicating how many of the parts of a unit are to be taken. In a common fraction, it appears above or to the left of the line.

C. A DENOMINATOR is the term of a fraction indicating the number of equal parts into which the unit is divided. In a common fraction, it appears below or to the right of the line.

D. EXAMPLES:

(1.) $\frac{1}{2} = \frac{\text{NUMERATOR}}{\text{DENOMINATOR}} = \text{FRACTION}$

(2.) NUMERATOR ⟶ 1/2 ⟵ DENOMINATOR

To Add or Subtract

TO SOLVE: $1/2 - 2/3 + 3/4 - 5/6 + 7/12 = ?$

A. Determine the lowest common denominator that each of the denominators 2, 3, 4, 6, and 12 will divide into an even number of times.

The lowest common denominator is 12.

B. Work one fraction at a time using the formula:

$$\frac{\textbf{COMMON DENOMINATOR}}{\textbf{DENOMINATOR OF FRACTION}} \times \textbf{NUMERATOR OF FRACTION}$$

(1.) $12/2 \times 1 = 6 \times 1 = 6$	$1/2$ becomes $6/12$
(2.) $12/3 \times 2 = 4 \times 2 = 8$	$2/3$ becomes $8/12$
(3.) $12/4 \times 3 = 3 \times 3 = 9$	$3/4$ becomes $9/12$
(4.) $12/6 \times 5 = 2 \times 5 = 10$	$5/6$ becomes $10/12$
(5.) $7/12$ remains $7/12$	

FRACTIONS

To Add or Subtract *(continued)*

C. We can now convert the problem from its original form to its new form using 12 as the common denominator.

$1/2 - 2/3 + 3/4 - 5/6 + 7/12$ = Original form

$\frac{6 - 8 + 9 - 10 + 7}{12}$ = Present form

$\frac{4}{12} = \frac{1}{3}$ Reduced to lowest form

D. To convert fractions to decimal form, simply divide the numerator of the fraction by the denominator of the fraction.

EXAMPLE: $\frac{1}{3}$ = 1 DIVIDED BY 3 = 0.333

To Multiply

A. The numerator of fraction #1 times the numerator of fraction #2 is equal to the numerator of the product.

B. The denominator of fraction #1 times the denominator of fraction #2 is equal to the denominator of the product.

C. EXAMPLE:

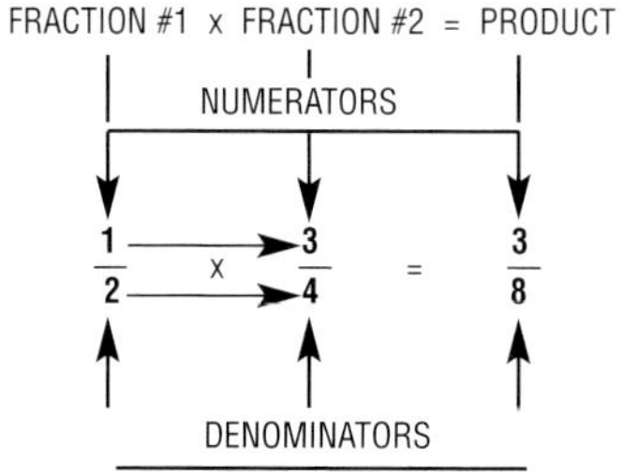

FRACTION #1 x FRACTION #2 = PRODUCT

NUMERATORS

$$\frac{1}{2} \times \frac{3}{4} = \frac{3}{8}$$

DENOMINATORS

NOTE: To change 3/8 to decimal form, divide 3 by 8 = 0.375

FRACTIONS

To Divide

A. The numerator of fraction #1 times the denominator of fraction #2 is equal to the numerator of the quotient.

B. The denominator of fraction #1 times the numerator of fraction #2 is equal to the denominator of the quotient.

C. EXAMPLE: $\frac{1}{2} \div \frac{3}{4}$

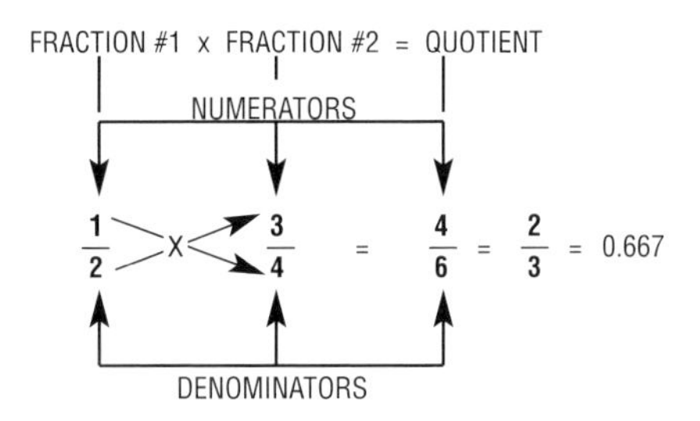

D. An alternate method for dividing by a fraction is to multiply by the reciprocal of the divisor (the second fraction in a division problem).

E. EXAMPLE: $\frac{1}{2} \div \frac{3}{4}$

The reciprocal of $\frac{3}{4}$ is $\frac{4}{3}$

so, $\frac{1}{2} \div \frac{3}{4} = \frac{1}{2} \times \frac{4}{3} = \frac{4}{6} = \frac{2}{3} = 0.667$

EQUATIONS

The word "EQUATION" means equal or the same as.

EXAMPLE: $2 \times 10 = 4 \times 5$

$$20 = 20$$

Rules

A. **The same number may be added to both sides of an equation without changing its values.**

EXAMPLE: $(2 \times 10) + 3 = (4 \times 5) + 3$

$$23 = 23$$

B. **The same number may be subtracted from both sides of an equation without changing its values.**

EXAMPLE: $(2 \times 10) - 3 = (4 \times 5) - 3$

$$17 = 17$$

C. **Both sides of an equation may be divided by the same number without changing its values.**

EXAMPLE: $\frac{2 \times 10}{20} = \frac{4 \times 5}{20}$

$$1 = 1$$

D. **Both sides of an equation may be multiplied by the same number without changing its values.**

EXAMPLE: $3 \times (2 \times 10) = 3 \times (4 \times 5)$

$$60 = 60$$

E. **TRANSPOSITION:**

The process of moving a quantity from one side of an equation to the other side of an equation by changing its sign of operation.

1. **A term may be transposed if its sign is changed from plus (+) to minus (–), or from minus (–) to plus (+).**

EXAMPLE: $X + 5 = 25$

$$X + 5 - \mathit{5} = 25 - \mathit{5}$$

$$X = 20$$

E. **TRANSPOSITION *(continued)*:**

2. **A multiplier may be removed from one side of an equation by making it a divisor on the other side; or a divisor may be removed from one side of an equation by making it a multiplier on the other side.**

EXAMPLE: Multiplier from one side of equation (4) becomes divisor on other side.

$$4X = 40 \text{ becomes } X = \frac{40}{4} = 10$$

EXAMPLE: Divisor from one side of equation becomes multiplier on other side.

$$\frac{X}{4} = 10 \text{ becomes } X = 10 \times 4$$

Signs

A. **ADDITION** of numbers with *DIFFERENT* signs:

1. **RULE: Use the sign of the larger and subtract.**

EXAMPLE:

$$\begin{array}{r} +3 \\ + -2 \\ \hline +1 \end{array} \qquad \begin{array}{r} -2 \\ + +3 \\ \hline +1 \end{array}$$

B. **ADDITION** of numbers with the *SAME* signs:

2. **RULE: Use the common sign and add.**

EXAMPLE:

$$\begin{array}{r} +3 \\ + +2 \\ \hline +5 \end{array} \qquad \begin{array}{r} -3 \\ + -2 \\ \hline -5 \end{array}$$

C. **SUBTRACTION** of numbers with *DIFFERENT* signs:

3. **RULE: Change the sign of the subtrahend (the second number in a subtraction problem) and proceed as in addition.**

EXAMPLE:

$$\begin{array}{r} +3 \\ - -2 \\ \hline \end{array} = \begin{array}{r} +3 \\ + +2 \\ \hline +5 \end{array} \qquad \begin{array}{r} -2 \\ - +3 \\ \hline \end{array} = \begin{array}{r} -2 \\ + -3 \\ \hline -5 \end{array}$$

EQUATIONS

Signs

D. **SUBTRACTION** of numbers with the *SAME* signs:

4. **RULE: Change the sign of the subtrahend (the second number in a subtraction problem) and proceed as in addition.**

EXAMPLE:

$$\begin{array}{r} +3 \\ -\ +2 \\ \hline \end{array} = \begin{array}{r} +3 \\ +\ -2 \\ \hline +1 \end{array} \qquad \begin{array}{r} -3 \\ -\ -2 \\ \hline \end{array} = \begin{array}{r} -3 \\ +\ +2 \\ \hline -1 \end{array}$$

E. **MULTIPLICATION:**

5. **RULE: The product of any two numbers having LIKE signs is POSITIVE. The product of any two numbers having UNLIKE signs is NEGATIVE.**

EXAMPLE:

$(+3) \times (-2) = -6$
$(-3) \times (+2) = -6$
$(+3) \times (+2) = +6$
$(-3) \times (-2) = +6$

F. **DIVISION:**

6. **RULE: If the divisor and the dividend have LIKE signs, the sign of the quotient is POSITIVE. If the divisor and dividend have UNLIKE signs, the sign of the quotient is NEGATIVE.**

EXAMPLE:

$\frac{+6}{-2} = -3 \qquad \frac{+6}{+2} = +3$

$\frac{-6}{+2} = -3 \qquad \frac{-6}{-2} = +3$

NATURAL TRIGONOMETRIC FUNCTIONS

ANGLE	SINE	COSINE	TANGENT	COTANGENT	SECANT	COSECANT	
0	.0000	1.0000	.0000		1.0000		90
1	.0175	.9998	.0175	57.2900	1.0002	57.2987	89
2	.0349	.9994	.0349	28.6363	1.0006	28.6537	88
3	.0523	.9986	.0524	19.0811	1.0014	19.1073	87
4	.0698	.9976	.0699	14.3007	1.0024	14.3356	86
5	.0872	.9962	.0875	11.4301	1.0038	11.4737	85
6	.1045	.9945	.1051	9.5144	1.0055	9.5668	84
7	.1219	.9925	.1228	8.1443	1.0075	8.2055	83
8	.1392	.9903	.1405	7.1154	1.0098	7.1853	82
9	.1564	.9877	.1584	6.3138	1.0125	6.3925	81
10	.1736	.9848	.1763	5.6713	1.0154	5.7588	80
11	.1908	.9816	.1944	5.1446	1.0187	5.2408	79
12	.2079	.9781	.2126	4.7046	1.0223	4.8097	78
13	.2250	.9744	.2309	4.3315	1.0263	4.4454	77
14	.2419	.9703	.2493	4.0108	1.0306	4.1336	76
15	.2588	.9659	.2679	3.7321	1.0353	3.8637	75
16	.2756	.9613	.2867	3.4874	1.0403	3.6280	74
17	.2924	.9563	.3057	3.2709	1.0457	3.4203	73
18	.3090	.9511	.3249	3.0777	1.0515	3.2361	72
19	.3256	.9455	.3443	2.9042	1.0576	3.0716	71
20	.3420	.9397	.3640	2.7475	1.0642	2.9238	70
21	.3584	.9336	.3839	2.6051	1.0711	2.7904	69
22	.3746	.9272	.4040	2.4751	1.0785	2.6695	68
23	.3907	.9205	.4245	2.3559	1.0864	2.5593	67
24	.4067	.9135	.4452	2.2460	1.0946	2.4586	66
25	.4226	.9063	.4663	2.1445	1.1034	2.3662	65
26	.4384	.8988	.4877	2.0503	1.1126	2.2812	64
27	.4540	.8910	.5095	1.9626	1.1223	2.2027	63
	COSINE	SINE	COTANGT.	TANGENT	COSECANT	SECANT	ANGLE

NATURAL TRIGONOMETRIC FUNCTIONS

ANGLE	SINE	COSINE	TANGENT	COTANGENT	SECANT	COSECANT	
28	.4695	.8829	.5317	1.8807	1.1326	2.1301	62
29	.4848	.8746	.5543	1.8040	1.1434	2.0627	61
30	.5000	.8660	.5774	1.7321	1.1547	2.0000	60
31	.5150	.8572	.6009	1.6643	1.1666	1.9416	59
32	.5299	.8480	.6249	1.6003	1.1792	1.8871	58
33	.5446	.8387	.6494	1.5399	1.1924	1.8361	57
34	.5592	.8290	.6745	1.4826	1.2062	1.7883	56
35	.5736	.8192	.7002	1.4281	1.2208	1.7434	55
36	.5878	.8090	.7265	1.3764	1.2361	1.7013	54
37	.6018	.7986	.7536	1.3270	1.2521	1.6616	53
38	.6157	.7880	.7813	1.2799	1.2690	1.6243	52
39	.6293	.7771	.8098	1.2349	1.2868	1.5890	51
40	.6428	.7660	.8391	1.1918	1.3054	1.5557	50
41	.6561	.7547	.8693	1.1504	1.3250	1.5243	49
42	.6691	.7431	.9004	1.1106	1.3456	1.4945	48
43	.6820	.7314	.9325	1.0724	1.3673	1.4663	47
44	.6947	.7193	.9657	1.0355	1.3902	1.4396	46
45	.7071	.7071	1.0000	1.0000	1.4142	1.4142	45
	COSINE	SINE	COTANGT.	TANGENT	COSECANT	SECANT	ANGLE

NOTE: For Angles 0 - 45, use Top Row & Left Column.

For Angles 45 - 90, use Bottom Row & Right Column.

TRIGONOMETRY

TRIGONOMETRY is the mathematics dealing with the relations of sides and angles of triangles.

A **TRIANGLE** is a figure enclosed by three straight sides. The sum of the three angles is 180 degrees. All triangles have six parts: three angles and three sides opposite the angles.

RIGHT TRIANGLES are triangles that have one angle of 90 degrees and two angles of less than 90 degrees.

To help you remember the six trigonometric functions, memorize:

"OH HELL ANOTHER HOUR OF ANDY"

SINE $\theta = \dfrac{\text{OPPOSITE SIDE } (OH)}{\text{HYPOTENUSE } (HELL)}$

COSINE $\theta = \dfrac{\text{ADJACENT SIDE } (ANOTHER)}{\text{HYPOTENUSE } (HOUR)}$

TANGENT $\theta = \dfrac{\text{OPPOSITE SIDE } (OF)}{\text{ADJACENT SIDE } (ANDY)}$

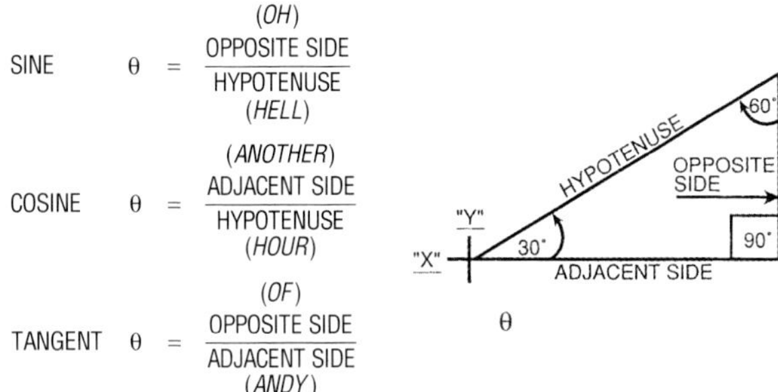

Now, use backward: **"ANDY OF HOUR ANOTHER HELL OH"**

COTANGENT $\theta = \dfrac{\text{ADJACENT SIDE } (ANDY)}{\text{OPPOSITE SIDE } (OF)}$

Always place the angle to be solved at the vertex (where "X" and "Y" cross).

SECANT $\theta = \dfrac{\text{HYPOTENUSE } (HOUR)}{\text{ADJACENT SIDE } (ANOTHER)}$

COSECANT $\theta = \dfrac{\text{HYPOTENUSE } (HELL)}{\text{OPPOSITE SIDE } (OH)}$

Note:

θ = Theta = Any Angle

BENDING OFFSETS WITH TRIGONOMETRY

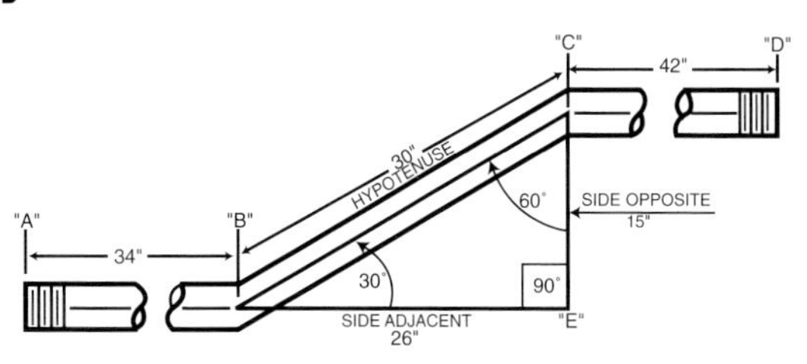

The Cosecant of the Angle Times the Offset Desired Is Equal to the Distance Between the Centers of the Bends.

Example:

To make a fifteen inch (15") offset, using thirty (30) degree bends:

1. Use Trig. Table (pages 154–155) to find the Cosecant of a thirty (30) degree angle. We find it to be two (2).
2. Multiply two (2) times the offset desired, which is fifteen (15) inches to determine the distance between bend "B" and bend "C". The answer is thirty (30) inches.

To mark the conduit for bending:

1. Measure from end of Conduit "A" thirty-four (34) inches to center of first bend "B", and mark.
2. Measure from mark "B" thirty (30) inches to center of second bend "C" and mark.
3. Measure from mark "C" forty-two (42) inches to "D", and mark. Cut, ream, and thread conduit before bending.

Rolling Offsets

To determine how much offset is needed to make a rolling offset:

1. Measure vertical required. Use work table (any square will do) and measure from corner this amount and mark.
2. Measure horizontal required. Measure ninety degrees from the vertical line measurement (starting in same corner) and mark.
3. The diagonal distance between these marks will be the amount of offset required.

NOTE: Shrink is hypotenuse minus the side adjacent.

ONE SHOT BENDS

SHRINK CONSTANT FOR ANGLES LESS THAN 60 DEGREES = Angle/120

EXAMPLE: The shrink constant for 45 degrees is 3/8"

45/120 = 3/8"

SHRINK CONSTANT FOR 60 TO 90 DEGREE ANGLES = Angle/100

EXAMPLE: The shrink constant for 45 degrees is 3/8"

45/100 = 3/8"

MULTIPLIER = (60/angle) + (Angle/200) - 0.15

EXAMPLE: The multiplier for 50 degrees is 1.3.

(60/50) + (50/200) - 0.15 = 1.3

The calculation for this multiplier is an error of less than half a percent.

BEND LENGTH = (Angle x D)/60

EXAMPLE: If putting a 40 degree bend in 3/4" conduit, the bend length is 4".

(40 x 6")/60 = 4"

"D" is the deduct for whatever size conduit is being run. This formula works for any angle between 0 and 90 degrees.

NOTE: With these formulas, the entire run in pieces (straight and curved) can be seen, including exactly where each piece starts and where it ends. This allows the bender direction (hook facing east or west) to be chosen at each point in the run, and bend marks can be laid out accordingly.

CHICAGO-TYPE BENDERS: 90° BENDING

"A" to "C" = STUB-UP
"C" to "D" = TAIL
"C" = BACK OF STUB-UP
"C" = BOTTOM OF CONDUIT

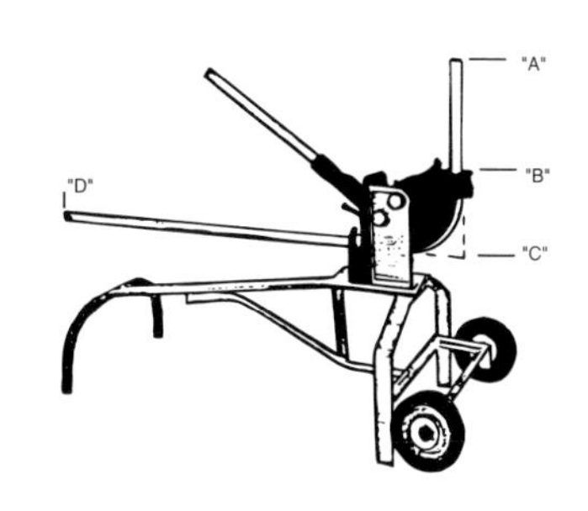

NOTE:
There are many variations of this type bender, but most manufacturers offer two sizes.
The *small* size shoe takes 1/2", 3/4", and 1" conduit.
The *large* size shoe takes $1\frac{1}{4}$" and $1\frac{1}{2}$" conduit.

TO DETERMINE THE "TAKE-UP" AND "SHRINK" OF EACH SIZE CONDUIT FOR A PARTICULAR BENDER TO MAKE NINETY DEGREE BENDS:

1. Use a straight piece of scrap conduit.
2. Measure exact length of scrap conduit, "A" to "D".

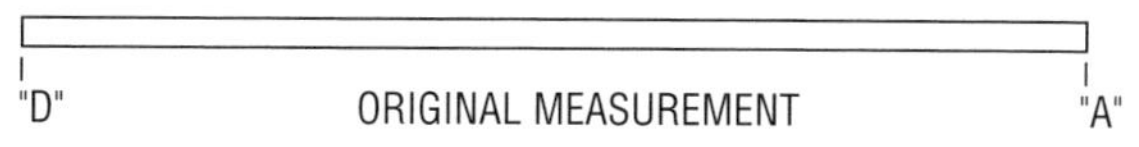

3. Place conduit in bender. Mark at edge of shoe, "B".
4. Level conduit. Bend ninety, and count number of pumps. Be sure to keep notes on each size conduit used.
5. After bending ninety:
   A. Distance between "B" and "C" is the TAKE-UP.
   B. Original measurement of the scrap piece of conduit subtracted from (distance "A" to "C" plus distance "C" to "D") is the SHRINK

NOTE: Both time and energy will be saved if conduit can be cut, reamed and threaded before bending. The same method can be used on hydraulic benders.

CHICAGO-TYPE BENDERS: OFFSETS

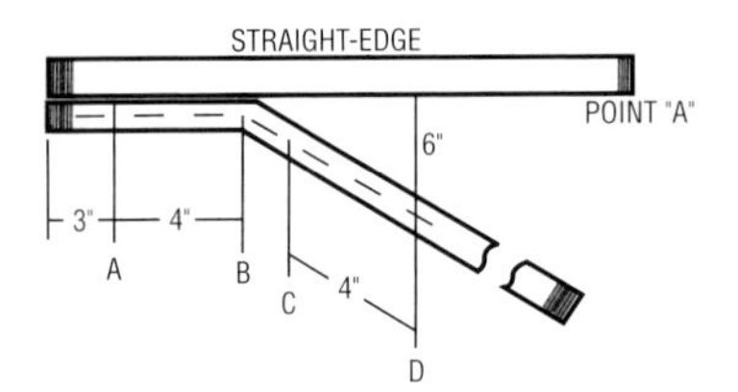

CHICAGO-TYPE BENDER

Example:

To bend a 6" offset:

1. Make a mark 3" from conduit end. Place conduit in bender with mark at outside edge of jaw.
2. Make three full pumps, making sure handle goes all the way down to the stop.
3. Remove conduit from bender and place alongside straight-edge.
4. Measure 6" from straight-edge to center of conduit. Mark point "D". Use square for accuracy.
5. Mark center of conduit from both directions through bend as shown by broken line. Where lines intersect is point "B".
6. Measure from "A" to "B" to determine distance from "D" to "C". Mark "C" and place conduit in bender with mark at outside edge of jaw, and with the kick pointing down. Use a level to prevent dogging conduit.
7. Make three full pumps, making sure handle goes all the way down to the stop.

NOTE: 1. There are several methods of bending rigid conduit with a Chicago-type bender, and any method that gets the job done in a minimal amount of time with craftsmanship is acceptable.

2. Whatever method is used, quality will improve with experience.

MULTI-SHOT: 90° CONDUIT BENDING

PROBLEM:

A. To measure, thread, cut, and ream conduit before bending.

B. To accurately bend conduit to the desired height of the stub-up (H) and to the desired length of the tail (L).

GIVEN:

A. Size of conduit = 2"

B. Space between conduit (center to center) = 6"

C. Height of stub-up = 36"

D. Length of tail = 48"

SOLUTION:

A. TO DETERMINE RADIUS (R):

Conduit #1 (inside conduit) will use the minimum radius unless otherwise specified. The minimum radius is eight times the size of the conduit. (see page 162)

RADIUS OF CONDUIT #1 = 8 x 2" + 1.25" = 17.25"
RADIUS OF CONDUIT #2 = RADIUS #1 + 6" = 23.25"
RADIUS OF CONDUIT #3 = RADIUS #2 + 6" = 29.25"

B. TO DETERMINE DEVELOPED LENGTH (DL): RADIUS X 1.57 = DL

DL OF CONDUIT #1 = R x 1.57 = 17.25" x 1.57 = 27"
DL OF CONDUIT #2 = R x 1.57 = 23.25" x 1.57 = 36.5"
DL OF CONDUIT #3 = R x 1.57 = 29.25" x 1.57 = 46"

C. TO DETERMINE LENGTH OF NIPPLE:

LENGTH OF NIPPLE, CONDUIT #1 = L + H + DL - 2R
= 48" + 36" + 27" - 34.5"
= 76.5"

LENGTH OF NIPPLE, CONDUIT #2 = L + H + DL - 2R
= 54" + 42" + 36.5" - 46.5"
= 86"

LENGTH OF NIPPLE, CONDUIT #3 = L + H + DL - 2R
= 60" + 48" + 46" - 58.5"
= 95.5"

NOTES: 1. For 90-degree bends, SHRINK = 2R - DL
2. For offset bends, SHRINK = HYPOTENUSE - SIDE ADJACENT

MULTI-SHOT: 90° CONDUIT BENDING

Layout and Bending:

A. To locate point "B", measure from point "A", the length of the stub-up minus the radius. On all three conduit, point "B" will be 18.75" from point "A". (see page 162).

B. To locate point "C", measure from point "D", the length minus the radius, (see page 162). On all three conduit, point "C" will be 30.75" from point "D". (see page 162).

C. Divide the developed length (point "B" to point "C") into equal spaces. Spaces should not be more than 1.75" to prevent wrinkling of the conduit. On Conduit #1, seventeen spaces of 1.5882" each would give us eighteen shots of 5 degrees each. Remember there is always one less space than shot. When determining the number of shots, choose a number that will divide into ninety an even number of times.

D. If an elastic numbered tape is not available, try the method illustrated.

A to B = Conduit #1
Developed Length = 27"

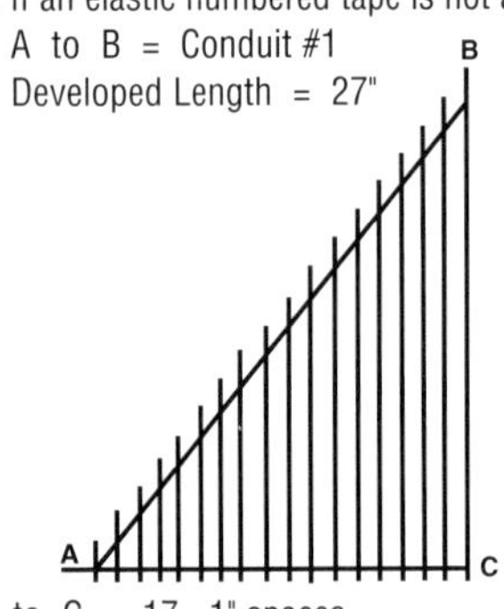

1.5882"
A
1"

A to C = 17 1" spaces
A to B = 17 1.5882" spaces
C = table or plywood corner

Measure from Point "C" (table corner) 17 inches along table edge to Point "A" and mark. Place end of rule at Point "A". Point "B" will be located where 27" mark meets table edge B-C. Mark on board, then transfer to conduit.

MULTI-SHOT: 90° CONDUIT BENDING

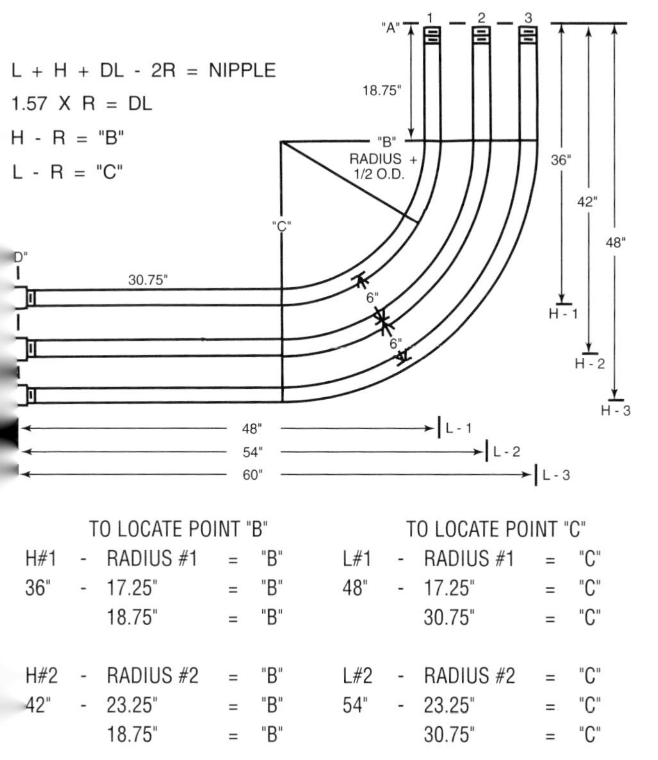

TO LOCATE POINT "B"					TO LOCATE POINT "C"				
H#1	-	RADIUS #1	=	"B"	L#1	-	RADIUS #1	=	"C"
36"	-	17.25"	=	"B"	48"	-	17.25"	=	"C"
		18.75"	=	"B"			30.75"	=	"C"
H#2	-	RADIUS #2	=	"B"	L#2	-	RADIUS #2	=	"C"
42"	-	23.25"	=	"B"	54"	-	23.25"	=	"C"
		18.75"	=	"B"			30.75"	=	"C"
H#3	-	RADIUS #3	=	"B"	L#3	-	RADIUS #3	=	"C"
48"	-	29.25"	=	"B"	60"	-	29.25"	=	"C"
		18.75"	=	"B"			30.75"	=	"C"

Points "B" and "C" are the same distance from the end on all three conduits.

OFFSET BENDS

EMT: Using Hand Bender

An offset bend is used to change the level, or plane, of the conduit. This is usually necessitated by the presence of an obstruction in the original conduit path.

Step One:

Determine the offset depth (X).

Step Two:

Multiply the offset depth "X" the multiplier for the degree of bend used to determine the distance between bends.

ANGLE		MULTIPLIER
10° x 10°	=	6
22½° x 22½°	=	2.6
30° x 30°	=	2
45° x 45°	=	1.4
60° x 60°	=	1.2

Example: If the offset depth required (X) is 6", and you intend to use 30° bends, the distance between bends is 6" x 2 = 12".

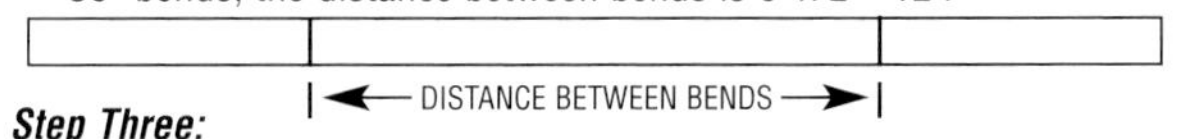

Step Three:

Mark at the appropriate points, align the arrow on the bender with the first mark, and bend to desired degree by aligning EMT with chosen degree line on bender.

Step Four:

Slide down the EMT, align the arrow with the second mark, and bend to the same degree line. Be sure to note the orientation of the bender head. Check alignment.

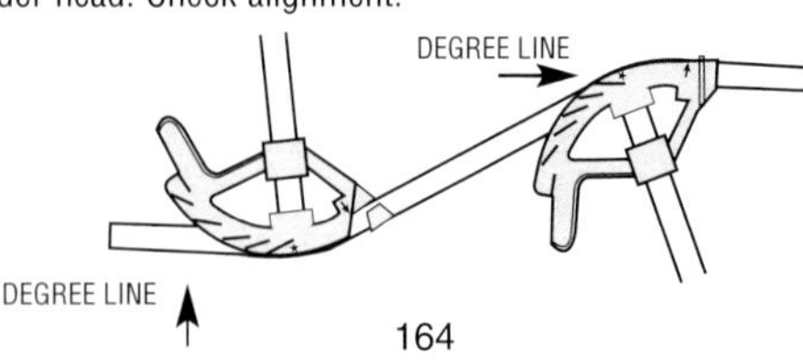

90° BENDS

EMT: Using Hand Bender

The stub-up is the most common bend.

Step One:

Determine the height of the stub-up required and mark on EMT.

Step Two:

Find the "Deduct" or "Take-up" amount from the Take-Up Chart. Subtract the take-up amount from the stub height and mark the EMT that distance from the end.

Step Three:

Align the arrow on bender with the last mark made on the EMT, and bend to the 90° mark on the bender.

DESCRIPTION		TAKE-UP
½" EMT	=	5"
¾" EMT	=	6"
1" EMT	=	8"
1¼" EMT	=	11"

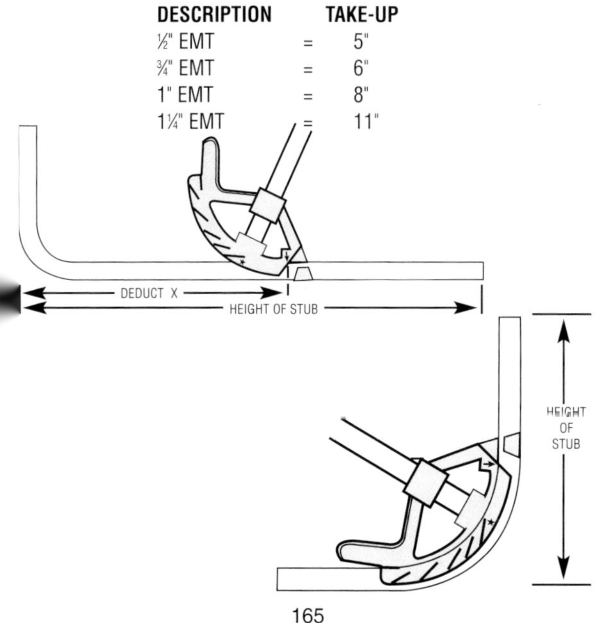

BACK-TO-BACK BENDS

EMT: Using Hand Bender

A back-to-back bend results in a "U" shape in a length of conduit. It's used for a conduit that runs along the floor or ceiling and turns up or down a wall.

Step One:

After the first 90° bend is made, determine the back-to-back length and mark on EMT.

Step Two:

Align this back-to-back mark with the star mark on the bender, and bend to 90°.

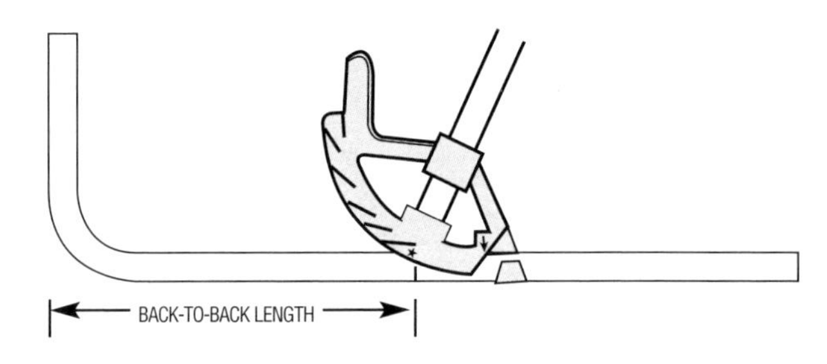

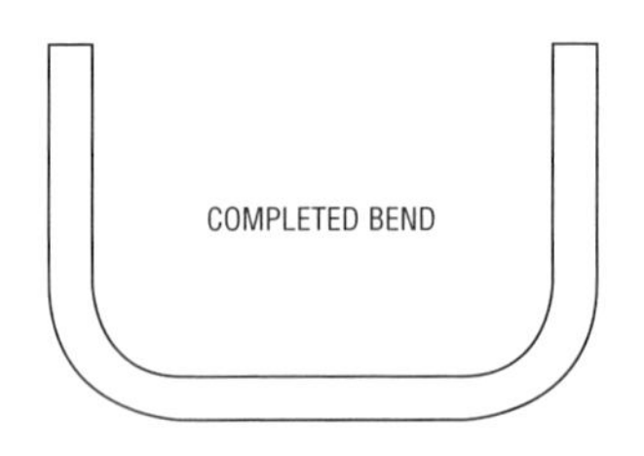

THREE-POINT SADDLE BENDS

EMT: Using Hand Bender

The 3-point saddle bend is used when encountering an obstacle (usually another pipe).

Step One:

Measure the height of the obstruction.
Mark the center point on EMT.

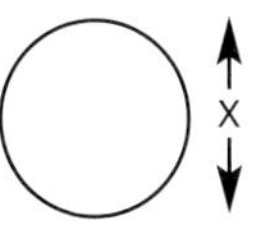

Step Two:

Multiply the height of the obstruction by 2.5 and mark this distance on each side of the center mark.

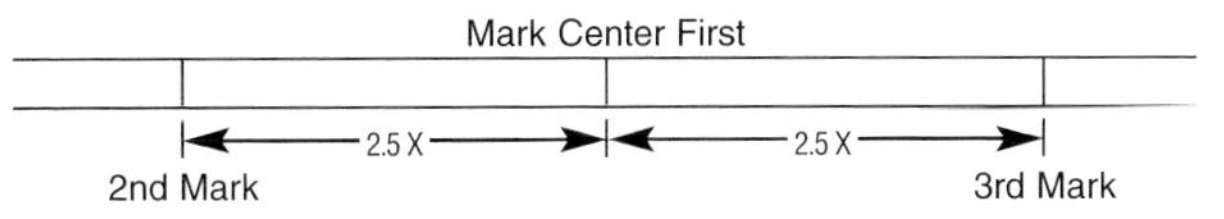

Step Three:

Place the center mark on the saddle mark or notch. Bend to 45°.

Step Four:

Bend the second mark to 22½° angle at arrow.

Step Five:

Bend the third mark to 22½° angle at arrow. Be aware of the orientation of the EMT on all bends. Check alignment.

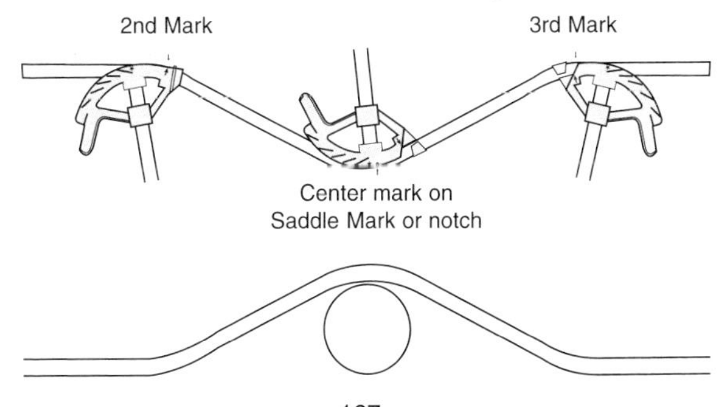

PULLEY CALCULATIONS

The most common configuration consists of a motor with a pulley attached to its shaft, connected by a belt to a second pulley. The motor pulley is referred to as the **Driving Pulley**. The second pulley is called the **Driven Pulley**. The speed at which the Driven Pulley turns is determined by the speed at which the Driving Pulley turns as well as the diameters of both pulleys. The following formulas may be used to determine the relationships between the motor, pulley diameters, and pulley speeds.

D = **Diameter of Driving Pulley**
d^1 = **Diameter of Driven Pulley**
S = **Speed of Driving Pulley** (revolutions per minute)
s^1 = **Speed of Driven Pulley** (revolutions per minute)

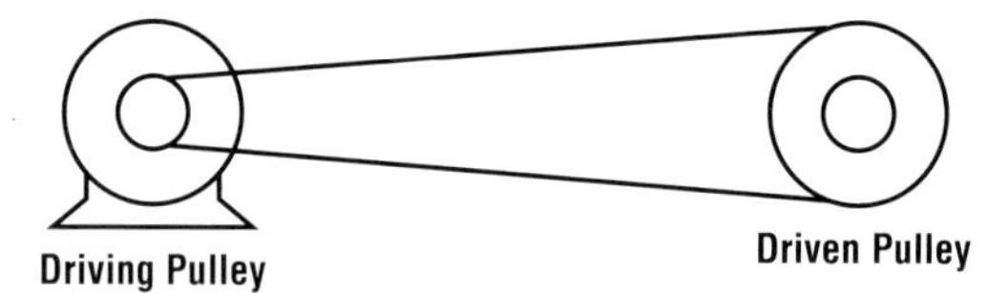

- *To determine the speed of the Driven Pulley (Driven RPM):*

$$s^1 = \frac{D \times S}{d^1} \quad \text{or} \quad \text{Driven RPM} = \frac{\text{Driving Pulley Dia.} \times \text{Driving RPM}}{\text{Driven Pulley Dia.}}$$

- *To determine the speed of the Driving Pulley (Driving RPM):*

$$S = \frac{d^1 \times s^1}{D} \quad \text{or} \quad \text{Driving RPM} = \frac{\text{Driven Pulley Dia.} \times \text{Driven RPM}}{\text{Driving Pulley Dia.}}$$

- *To determine the diameter of the Driven Pulley (Driven Dia.):*

$$d^1 = \frac{D \times S}{s^1} \quad \text{or} \quad \text{Driven Dia.} = \frac{\text{Driving Pulley Dia.} \times \text{Driving RPM}}{\text{Driven RPM}}$$

- *To determine the diameter of the Driving Pulley (Driving Dia.):*

$$D = \frac{d^1 \times s^1}{S} \quad \text{or} \quad \text{Driving Dia.} = \frac{\text{Driven Pulley Dia.} \times \text{Driven RPM}}{\text{Driving RPM}}$$

USEFUL KNOTS

BOWLINE

RUNNING BOWLINE

BOWLINE ON THE BIGHT

CLOVE HITCH

SHEEPSHANK

ROLLING HITCH

SINGLE BLACKWALL HITCH

CATSPAW

DOUBLE BLACKWALL HITCH

SQUARE KNOT

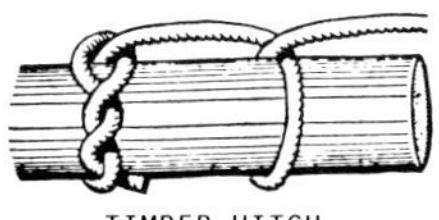

TIMBER HITCH WITH HALF HITCH

SINGLE SHEET BEND

HAND SIGNALS

STOP

DOG EVERYTHING

EMERGENCY STOP

TRAVEL

TRAVEL BOTH TRACKS (CRAWLER CRANES ONLY)

TRAVEL ONE TRACK (CRAWLERS)

RETRACT BOOM

EXTEND BOOM

SWING BOOM

HAND SIGNALS

RAISE LOAD

LOWER LOAD

MAIN HOIST

MOVE SLOWLY

RAISE BOOM AND LOWER LOAD (FLEX FINGERS)

LOWER BOOM AND RAISE LOAD (FLEX FINGERS)

USE WHIP LINE

BOOM UP

BOOM DOWN

ELECTRICAL SAFETY DEFINITIONS

(Courtesy of Littelfuse® POWR-GARD® Products)

Arc-Flash - The sudden release of heat energy and intense light at the point of an arc. Can be considered a short circuit through the air, usually created by accidental contact between live conductors.

Arc-Blast - A pressure wave created by the heating, melting, vaporization, and expansion of conducting material and surrounding gases of air.

Arc Gap - The distance between energized conductors or between energized conductors and ground. Shorter arc gaps result in less energy being expended in the arc, while longer gaps reduce arc current. For 600 volts and below, arc gaps of 1.25 inches (32 mm) typically produce the maximum incident energy.

Approach Boundaries - Protection boundaries established to protect personnel from shock.

Calorie - The amount of heat needed to raise the temperature of 1 gram of water by 1 degree Celsius. One cal/cm^2 is equivalent to the exposure on the tip of a finger by a cigarette lighter for 1 second.

Distance to Arc - Refers to the distance from the receiving surface to the arc center. The value used for most calculations is typically 18 inches.

Electrically Safe Work Condition - Condition where the equipment and or circuit components have been disconnected from electrical energy sources, locked/tagged out, and tested to verify all sources of power are removed.

Exposed Live Parts - An energized conductor or part that is capable of being inadvertently touched or approached (nearer than a safe distance) by a person. It is applicable to parts that are not in an electrically safe work condition, suitably grounded, isolated, or insulated.

Flame Resistant (FR) - A term referring to fabric and its ability to limit the ignition or burning of the garment. It can be a specific characteristic of the material or a treatment applied to a material.

For more information, refer to *NFPA 70E®, Standard for Electrical Safety in the Workplace.*

ELECTRICAL SAFETY DEFINITIONS

(Courtesy of Littelfuse® POWR-GARD® Products)

Flash Hazard Analysis - A study that analyzes potential exposure to Arc-Flash hazards. The outcome of the study establishes Incident Energy levels, Hazard Risk Categories, Flash Protection Boundaries, and required PPE. It also helps define safe work practices.

Flash Protection Boundary - A protection boundary established to protect personnel from Arc-Flash hazards. The Flash Protection Boundary is the distance at which an unprotected worker can receive a second-degree burn to bare skin.

Flash Suit - A term referring to a complete FR rated Personal Protective Equipment (PPE) system that would cover a person's body, excluding the hands and feet. Included would be pants, shirt/jacket, and flash hood with a built-in face shield.

Hazard Risk Category - A classification of risks (from 0-4) defined by NFPA 70E. Each category requires PPE and is related to incident energy levels.

Incident Energy - The amount of thermal energy impressed on a surface generated during an electrical arc at a certain distance from the arc. Typically measured in cal/cm^2.

PPE - An acronym for Personal Protective Equipment. It can include clothing, tools and equipment.

Qualified Person - A person who is trained and knowledgeable on the construction and operation of the equipment and can recognize and avoid electrical hazards that may be encountered.

Unqualified Person - A person that does not possess all the skills and knowledge or has not been trained for a particular task.

Shock - A trauma subjected to the body by electrical current. When personnel come in contact with energized conductors, it can result in current flowing through their body often causing serious injury or death.

For more information, refer to *NFPA 70E®, Standard for Electrical Safety in the Workplace.*

ELECTRICAL SAFETY CHECKLIST

(Courtesy of Littelfuse® POWR-GARD® Products)

1. De-energize the equipment whenever possible prior to performing any work.
2. Verify you are "qualified" and properly trained to perform the required task.
3. Identify the equipment and verify you have a clear understanding and have been trained on how the equipment operates.
4. Provide justification why the work must be performed in an "energized" condition (if applicable).
5. Identify which safe work practices will be used to insure safety.
6. Determine if a Hazard Analysis has been performed to identify all hazards (Shock, Arc-Flash, etc.).
7. Identify protection boundaries for Shock (Limited, Restricted, & Prohibited Approach) and Arc-Flash (Flash Protection Boundary).
8. Identify the required Personal Protective Equipment (PPE) for the task to be performed based on the Hazard Risk Category (HRC) and available incident Energy (cal/cm^2).
9. Provide barriers or other means to prevent access to the work area by "unqualified" workers.
10. Perform a job briefing and identify job or task specific hazards.
11. Obtain written management approval to perform the work in an "energized" condition (where applicable).

For more information, refer to *NFPA 70E®*, *Standard for Electrical Safety in the Workplace.*

ELECTRICAL SAFETY: LOCKOUT–TAGOUT PROCEDURES

(Courtesy of Littelfuse® POWR-GARD® Products)

OSHA requires that energy sources to machines or equipment must be turned off and disconnected isolating them from the energy source. The isolating or disconnecting means they must be either locked or tagged with a warning label. While lockout is the more reliable and preferred method, OSHA accepts tagout to be a suitable replacement in limited situations. NFPA 70E Article 120 contains detailed instructions for lockout/tagout and placing equipment in an Electrically Safe Work Condition.

Application of Lockout–Tagout Devices

1. Make necessary preparations for shutdown.
2. Shut down the machine or equipment.
3. Turn OFF (open) the energy isolating device (fuse/circuit breaker).
4. Apply the lockout or tagout device.
5. Render safe all stored or residual energy.
6. Verify the isolation and deenergization of the machine or equipment.

Removal of Lockout–Tagout Devices

1. Inspect the work area to ensure that nonessential items have been removed and that machine or equipment components are intact and capable of operating properly. Especially look for tools or pieces of conductors that may have not been removed.
2. Check the area around the machine or equipment to ensure that all employees have been safely positioned or removed.
3. Make sure that only the employees who attached the locks or tags are the ones that are removing them
4. After removing locks or tags, notify affected employees before starting equipment or machines.

'OTE: For specific Lockout–Tagout procedures, refer to OSHA and NFPA 70E.

ELECTRICAL SAFETY: SHOCK PROTECTION BOUNDARIES

(Courtesy of Littelfuse® POWR-GARD® Products)

Nominal System Voltage (Phase to Phase)	Limited Approach Boundary Exposed Movable Conductor	Limited Approach Boundary Exposed Fixed Circuit Part	Restricted Approach Boundary	Prohibited Approach Boundary
50 to 300 V	10 ft. 0 in.	3 ft. 6 in.	Avoid Contact	Avoid Contac
301 to 750 V	10 ft. 0 in.	3 ft. 6 in.	1 ft. 0 in.	0 ft. 1 in.
751 V to 15 kV	10 ft. 0 in.	5 ft. 0 in.	2 ft. 2 in.	0 ft. 7 in.
15.1 kV to 36 kV	10 ft. 0 in.	6 ft. 0 in.	2 ft. 7 in.	0 ft. 10 in.
36.1 kV to 46 kV	10 ft. 0 in.	8 ft. 0 in.	2 ft. 9 in.	1 ft. 5 in.
46.1 kV to 72.5 kV	10 ft. 0 in.	8 ft. 0 in.	3 ft. 2 in.	2 ft. 2 in.
72.6 kV to 121 kV	10 ft. 8 in.	8 ft. 0 in.	3 ft. 3 in.	2 ft. 9 in.

NOTE: Data derived from NFPA 70E Table 130.2(C).

Shock protection boundaries are based on system voltage and whether the exposed energized components are fixed or movable. NFPA 70E Table 130.2(C) defines these boundary distances for nominal phase-to-phase system voltages from 50 Volts to 800 kV. Approach Boundary distances may range from an inch to several feet. Please refer to NFPA 70E Table 130.2(C) for more information.

Protection Boundaries

Limited Approach: Qualified person or unqualified person if accompanied by qualified person. PPE is required.

Restricted Approach: Qualified persons only. PPE is required.

Prohibited Approach: Qualified persons only. Use PPE as if making direct contac with a live part.

ELECTRICAL SAFETY: HOW TO READ A WARNING LABEL

(Courtesy of Littelfuse® POWR-GARD® Products)

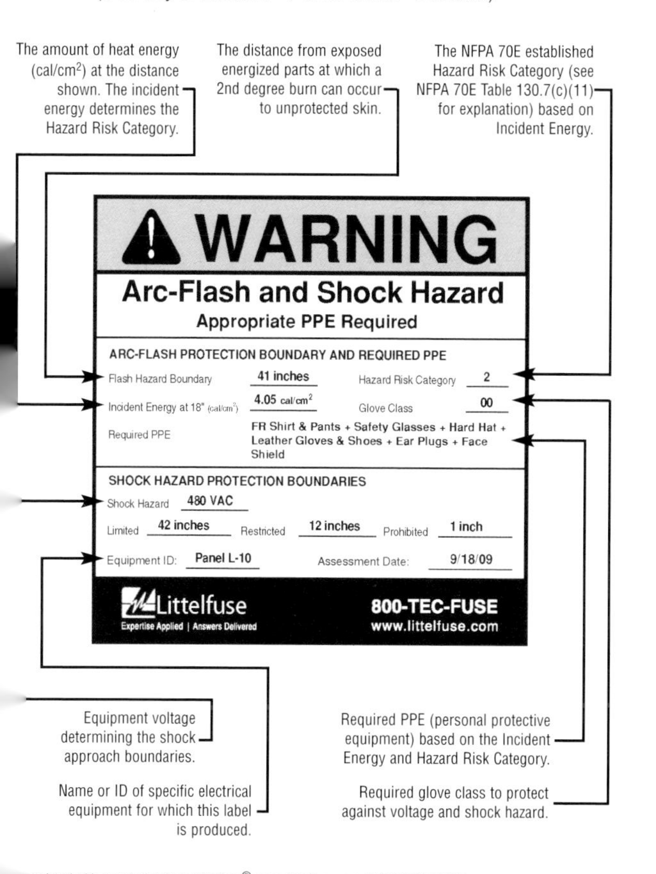

eprinted with permission from Littelfuse®; www.littelfuse.com; 1-800-TEC-FUSE

ELECTRICAL SAFETY: PERSONAL PROTECTION EQUIPMENT GUIDE

(Courtesy of Littelfuse® POWR-GARD® Products)

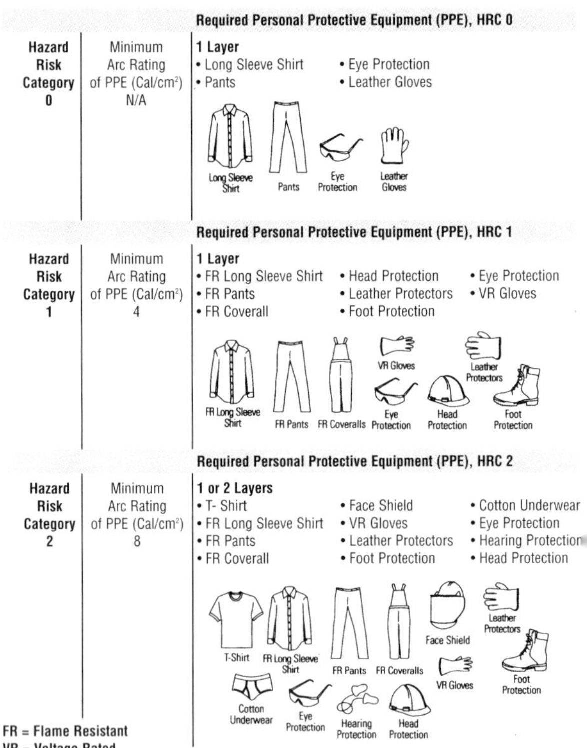

Hazard Risk Category	Minimum Arc Rating of PPE (Cal/cm^2)	Required Personal Protective Equipment (PPE)
0	N/A	**Required Personal Protective Equipment (PPE), HRC 0** **1 Layer** • Long Sleeve Shirt • Pants • Eye Protection • Leather Gloves
1	4	**Required Personal Protective Equipment (PPE), HRC 1** **1 Layer** • FR Long Sleeve Shirt • FR Pants • FR Coverall • Head Protection • Leather Protectors • Foot Protection • Eye Protection • VR Gloves
2	8	**Required Personal Protective Equipment (PPE), HRC 2** **1 or 2 Layers** • T- Shirt • FR Long Sleeve Shirt • FR Pants • FR Coverall • Face Shield • VR Gloves • Leather Protectors • Foot Protection • Cotton Underwear • Eye Protection • Hearing Protection • Head Protection

FR = Flame Resistant
VR = Voltage Rated

Disclaimer: The table above is for illustrative purposes only. PPE may vary depending on specific task. For additional information on electrical safety, see NFPA 70E®, *Standard for Electrical Safety in the Workplace* or NFPA 70E *Handbook for Electrical Safety in the Workplace*.

ELECTRICAL SAFETY: PERSONAL PROTECTION EQUIPMENT GUIDE

(Courtesy of Littelfuse® POWR-GARD® Products)

		Required Personal Protective Equipment (PPE), HRC 3
Hazard Risk Category 3	Minimum Arc Rating of PPE (Cal/cm^2) 25	**2 or 3 Layers** • T- Shirt • FR Long Sleeve Shirt • FR Pants • FR Coverall • Flash Hood • VR Gloves • Leather Protectors • Foot Protection • Cotton Underwear • Eye Protection • Hearing Protection • Head Protection
		Required Personal Protective Equipment (PPE), HRC 4
Hazard Risk Category 4	Minimum Arc Rating of PPE (Cal/cm^2) 40	**3 or more Layers** • T- Shirt • FR Long Sleeve Shirt • FR Pants • Multi-Layer Flash Suit • Flash Hood • VR Gloves • Leather Protectors • Foot Protection • Cotton Underwear • Eye Protection • Hearing Protection • Head Protection

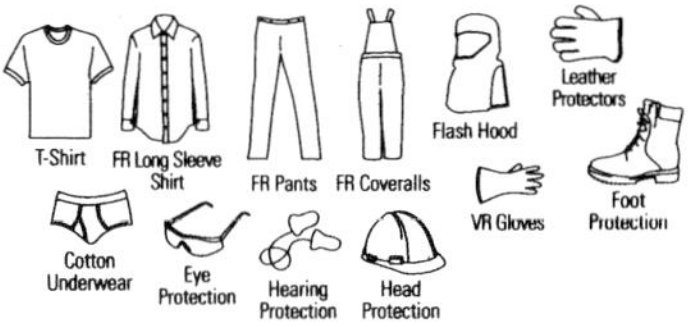

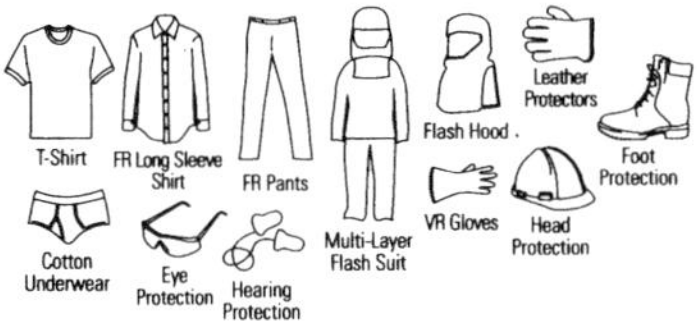

FR = Flame Resistant
VR = Voltage Rated

Disclaimer: The table above is for illustrative purposes only. PPE may vary depending on specific task. For additional information on electrical safety, see NFPA 70E®, *Standard for Electrical Safety in the Workplace* or NFPA 70E *Handbook for Electrical Safety in the Workplace*.

ALTERNATIVE ENERGY

Distributed generation systems are designed to work either independently or in parallel with the electric utility grid and have the goal of reducing utility billing, improving electrical reliability, or selling power back to the utility, and being less harmful to the environment. There are five basic types of distributed generation systems: engine-generation systems, solar photovoltaic systems, wind turbines, fuel cells, and microturbines.

Engine-Generation Systems

Engine-generation is the most common type of distributed generation system currently available and can be used almost anywhere. Engine-generators have the following:

- An internal combustion engine that runs on a variety of fuels.
- Components that consist of the engine and either an induction generator or a synchronous generator.
- An engine that is either a standby rated or a prime rated. A standby rated engine is rated to deliver power for the duration of a utility outage. A prime-rated engine is rated to deliver a continuous output with approximately 10% reserved for surges.

Solar Photovoltaic Systems

Solar photovoltaic power converts sunlight to dc electrical energy. Solar is limited because of its requirement of sunlight.

- The operation of a solar system is automatic.
- Components consist of foundation and supports, either fixed or tracking arrays, and one or more inverters.
- Per the *NEC* 690.4(E), solar-associated wiring shall be installed by qualified persons only.

ALTERNATIVE ENERGY

Wind Turbines

Wind power converts wind to either ac or dc electrical energy. Wind is limited because it needs to be in an area of steady reliable wind.

- Wind is useful as a supplemental power source, but not as a backup source.
- The components of wind power are self-contained wind turbines and support towers.
- The turbine generator can be either directly connected to the fan blade or by a gearbox.
- See *NEC* Article 694 for more information regarding wind turbine systems.

ALTERNATIVE ENERGY

Fuel Cells

Fuel cells use an electrochemical reactor to generate dc electrical energy. Fuel cells are basically batteries that use hydrogen and oxygen as fuel instead of storing electrical energy.

- A fuel cell generator has no moving parts.
- The fuel cell is composed of a fuel processor, individual fuel cells, fuel cell stack, and power-conditioning equipment.
- Fuel cells can have extremely high operating temperatures, which can limit where they can be used.
- The fuel processor converts hydrocarbon fuel into a relatively pure hydrogen gas.
- Fuel cell systems cannot be installed in a Class I hazardous locations (see *NEC* Article 501) and should be installed outside where possible. See *NEC* Article 692 for more information on fuel cell systems. See *NEC* 700.12(E), 701.1(F), 705.3(C), and 708.20(H) for additional information regarding fuel cell systems and their requirements.

Microturbines

Microturbines are small, single-staged combustion turbines. They can generate either ac or dc electrical energy. Microturbines are also limited in where they can be used due to their high operating temperatures.

- Microturbines range in size from 25 to 500 kW and are modular. They can operate on a wide variety of fuels but are only considered as a renewable energy source.
- The components of microturbines are the compressor, combustion chamber, turbine, generator, recuperator, and the power controller.
- Microturbines are capable of being a stand-alone unit, but the generator loading needs to be relatively steady due to the microturbines' inability to respond quickly.

nterconnected Generation Systems

nterconnected generation systems are one of two basic types: passive or active. Passive generation technologies have no control over power production (wind and solar). Active generation technologies have control over power production and can be regulated to load demands. All grid-connected generation systems must comply with *NEC* Article 705 and with IEEE 1547, *Standard for Interconnecting Distributed Resources with Electric Power Systems.*

Utility-interactive power inverters regulate the conversion of dc power into 60 Hz ac voltage waveform in parallel with another ac source (e.g., the electric grid). These systems should comply with *NEC* Articles 690, 692, and 705, as required.

Distributed generation systems that are capable of being connected to the grid must have a disconnecting means capable of disconnecting from the grid to prevent the potential hazard of back-feed. See *NEC* Article 404.6(C).

FIRST AID

The first line of defense is turning off the power and establishing an electrically safe work condition. In addition, the best protection from the consequences of severe injury or illness is the knowledge of first aid practices and the ability to act in an emergency situation. The Emergency Care and Safety Institute is an educational organization created for the purpose of delivering the highest quality training to lay people and professionals in the areas of first aid, CPR, AED, bloodborne pathogens, and related safety and health areas. The content of the training materials used by the Emergency Care and Safety Institute is approved by the American Academy of Orthopaedic Surgeons (AAOS) and the American College of Emergency Physicians (ACEP), two of the most respected names in injury, illness, and emergency medical care.

Visit www.ECSInstitute.org for more information.

Scene Size-Up

When approaching the scene of an emergency, take a few seconds to size up the scene to assess the following:

- Danger to the rescuer and to the victim. Scan the area for immediate dangers to yourself or to the victim. If the scene is unsafe, make it safe. If you are unable to do so, do not enter.
- Type of problem—injury or illness. This helps to identify what is wrong.
- Number of victims. Determine how many people are involved. There may be more than one person, so look around and ask about others.

If there are two or more victims, first check those who are not moving or talking. These are the individuals who may need your help first.

FIRST AID

How to Call for Help

To receive emergency assistance of every kind in most communities, simply call 9-1-1. At some government installations and industrial sites, an additional system may apply. This should be an element of a job briefing. In any case, be prepared to tell the emergency medical services (EMS) dispatcher the following:

- Your name and phone number
- Exact location or address of emergency
- What happened
- Number of people
- Victim's condition and what is being done for the victim

Do not hang up until the dispatcher hangs up—the EMS dispatcher may be able to tell you how to care for the victim until the ambulance arrives.

Airway Obstruction

Management of Responsive Choking Victim

1. Check the victim for choking by asking, "Are you choking?" A choking person is unable to breathe, talk, cry, or cough.
2. Have someone call 9-1-1.
3. Position yourself behind the victim and locate the victim's navel.
4. Place a fist with the thumb side against the victim's abdomen just above the navel, grasp it with the other hand, and press it into the victim's abdomen with quick inward and upward thrusts. Continue thrusts until the object is removed or the victim becomes unresponsive.

If the victim becomes unresponsive, call 9-1-1 and give CPR.

FIRST AID

Adult Cardiopulmonary Resuscitation (CPR)

1. Check responsiveness by tapping the victim and asking, "Are you okay?"

2. At the same time, check for breathing by looking for chest rise and fall. If the victim is unresponsive and not breathing, he or she needs CPR (Step 4).

3. Have someone call 9-1-1 and have someone else retrieve an AED if available.

4. Perform CPR.
   - Place the heel of one hand on the center of the chest between the nipples. Place the other hand on top of the first hand.
   - Depress the chest 2 inches.
   - Give 30 chest compressions at a rate of at least 100 compressions per minute, allowing the chest to return to its normal position after each compression.
   - Tilt the head back and lift the chin. Pinch the nose, and give 2 breaths (1 second each).

5. Continue cycles of 30 chest compressions and two breaths until an AED is available, the victim shows signs of life, EMS takes over, or you are too tired to continue.

Bleeding

1. For a shallow wound, wash it with soap and water and flush with running water. Apply an antibiotic and cover the wound with a clean dressing.

2. For a deep wound, do not attempt to clean the wound; just stop the bleeding. Deep wounds require cleaning by a medically trained person. Cover large, gaping wounds with sterile gauze pads and apply pressure to stop the bleeding. Secure the gauze pad snugly with a bandage.

FIRST AID

Protect yourself against diseases carried by blood by wearing disposable medical exam gloves, using several layers of cloth or gauze pads, using waterproof material such as plastic, or having the victim apply pressure using his or her own hand.

Burns

Care for Burns

1. Stop the burning! Use water or smother flames.
2. Cool the burn. Apply cool water or cool, wet cloths until pain decreases (usually within 10 minutes).
3. Apply aloe-vera gel on first-degree burns (skin turns red). Apply antibiotic ointment on second-degree burns (skin blisters). Apply non-stick dressing on second- and third-degree burns (full thickness; penetrates skin layers, muscle, and fat). Seek medical attention if any of these conditions exist:
   - Breathing difficulty
   - Head, hands, feet, or genitals involved
   - Victim is elderly or very young
   - Involves electricity or chemical exposure
   - Second-degree burns cover more than an area equivalent to size of victim's entire back or chest
   - Any third-degree burns

Electrical Burns

- Check the scene for electrical hazards.
- Check breathing; CPR may be needed.

Frostbite

Recognizing Frostbite

The signs of frostbite include the following:

- White, waxy-looking skin
- Skin feels cold and numb (pain at first, followed by numbness)
- Blisters, which may appear after rewarming

Care for Frostbite

1. Move the victim to a warm place.
2. Remove any wet/cold clothing and any jewelry from the affected part.
3. Seek medical care.

Heart Attack

Recognizing a Heart Attack

Prompt medical care at the onset of a heart attack is vital to survival and the quality of recovery. This is sometimes easier said than done because many victims deny they are experiencing something as serious as a heart attack. The signs of a heart attack include the following:

- Chest pressure, squeezing, or pain lasting more than a few minutes. It may come and go.
- Pain spreading to either shoulder, the neck, the lower jaw, or either arm.
- Any or all of the following: weakness, dizziness, sweating, nausea, or shortness of breath.

Care for a Heart Attack

1. Seek medical care by calling 9-1-1. Medications to dissolve a clot are available but must be given early.
2. Help the victim into the most comfortable resting position.
3. If the victim is alert, able to swallow, and not allergic to aspirin, give one adult aspirin or four chewable aspirin.
4. If the victim has been prescribed medication for heart disease, such as nitroglycerin, help the victim to use it.
5. Monitor the victim's breathing.

Heat Cramps

Recognizing Heat Cramps

The signs of heat cramps include the following:

- Painful muscle spasms during or after physical activity

Care for Heat Cramps

1. Have the victim stop activity and rest in a cool area.
2. Stretch the cramped muscle.
3. If the victim is responsive and not nauseated, provide water or a commercial sport drink (such as Gatorade® or Powerade®).

Heat Exhaustion

Recognizing Heat Exhaustion

The signs of heat exhaustion can include the following:

- Heavy sweating
- Severe thirst

- Weakness
- Headache
- Nausea and vomiting

Care for Heat Exhaustion

1. Have the victim stop activity and rest in a cool area.
2. Remove any excess or tight clothing.
3. If the victim is responsive and not nauseated, provide water or a commercial sport drink (such as Gatorade® or Powerade®).
4. Have the victim lie down.
5. Apply cool packs to the armpits and to the crease where the legs attach to the pelvis.

Seek medical care if the condition does not improve within 30 minutes.

Heatstroke

Recognizing Heatstroke

The signs of heatstroke can include the following:

- Extremely hot skin
- Dry skin (may be wet at first)
- Confusion
- Seizures
- Unresponsiveness

Care for Heatstroke

1. Have the victim stop activity and rest in a cool area.
2. Call 9-1-1.
3. If the victim is unresponsive and not breathing, begin CPR.
4. Rapidly cool the victim by whatever means possible: cool, wet towels or sheets to the head and body accompanied by fanning, and/or cold packs against the armpits, the sides of the neck, and the groin.

Hypothermia

Recognizing Hypothermia

The signs of hypothermia include the following:

- Uncontrollable shivering
- Confusion, sluggishness
- Cold skin (even under clothing)

Care for Hypothermia

1. Get the victim out of the cold.
2. Prevent heat loss by:
    - replacing wet clothing with dry clothing,
    - covering the victim's head, and
    - placing insulation (such as blankets, towels, coats) beneath and over the victim.
3. Have the victim rest in a comfortable position.
4. If the victim is alert and able to swallow, give him or her warm, sugary beverages.
5. Seek medical care for severe hypothermia (rigid muscles, cold skin on abdomen, confusion, or lethargy).

Ingested Poisons

Recognizing Ingested Poisoning

The signs of ingested poisoning include the following:

- Abdominal pain and cramping
- Nausea or vomiting
- Diarrhea
- Burns, odor, or stains around and in the mouth
- Drowsiness or unresponsiveness
- Poison container nearby

Care for Ingested Poisoning

1. Try to determine what poison was swallowed and how much. Also make note of the age and size of the victim.
2. For a responsive patient, call the poison control center for instructions: 800-222-1222. For an unresponsive victim, call 9-1-1.
3. Place the victim on his or her side if vomiting occurs.

Shock

Recognizing Shock

The signs of shock include the following:

- Altered mental status (agitation, anxiety, restlessness, and confusion)
- Pale, cold, and clammy skin, lips, and nail beds
- Nausea and vomiting
- Rapid breathing
- Unresponsiveness (when shock is severe)

Care for Shock

Even if there are no signs of shock, you should still treat seriously injured or suddenly ill victims for shock.

1. Place the victim on his or her back.
2. Place blankets under and over the victim to keep the victim warm.
3. Call 9-1-1.

Stroke

Recognizing a Stroke

The signs of a stroke include the following:

- Sudden weakness or numbness of the face, an arm, or a leg on one side of the body
- Blurred or decreased vision, especially on one side of the visual field
- Problems speaking
- Dizziness or loss of balance
- Sudden, severe headache

Care for a Stroke

1. Call 9-1-1.
2. If the victim is responsive, have the victim rest in the most comfortable position. This is often on his or her back with the head and shoulders slightly elevated.
3. If vomiting, place the victim on his or her side to keep the airway clear.